DRAINAGE CHARACTERISTICS and FLOODING PROBLEMS in AHOADA WEST/EAST and OGBA-EGBEMA NDONI LOCAL GOVERNMENT AREAS of RIVERS STATE, NIGERIA

Chukudi V. Izeogu, Ph.D.

Drainage Characteristics and Flooding Problems In Ahoada West/East and Ogba-Egbema Ndoni Local Government Areas of Orashi-Sombreiro Plains of Rivers State, Nigeria
All Rights Reserved.
Copyright © 2021 Chukudi V. Izeogu
v2.0

The opinions expressed in this manuscript are solely the opinions of the author and do not represent the opinions or thoughts of the publisher. The author has represented and warranted full ownership and/or legal right to publish all the materials in this book.

This book may not be reproduced, transmitted, or stored in whole or in part by any means, including graphic, electronic, or mechanical without the express written consent of the publisher except in the case of brief quotations embodied in critical articles and reviews.

Outskirts Press, Inc.
http://www.outskirtspress.com

ISBN: 978-1-9772-4513-7

Cover Photo © 2021 Chukudi V. Izeogu. All rights reserved - used with permission.

Outskirts Press and the "OP" logo are trademarks belonging to Outskirts Press, Inc.

PRINTED IN THE UNITED STATES OF AMERICA

Table of Contents

Synopsis i

Acknowledgement iii

Introduction v

1. The Environmental Background of the Ahoada Area
   in the Central Orashi–Sombreiro Plains 1
   The Study Area 1
   Land Forms and Climatic Factors 4
   Stratigraphical and Structural Characteristics of the
       Orashi-Sombreiro Plains 6
   The Major Drainage Systems in the Ahoada East and
       West and Ogba-Egbema Areas of Orashi-Sombreiro Plains 8

2. Some Aspect of Hydraulic Geometry of the Main Drainage
   Systems of the Studyarea: Orashi and Sombreiro Rivers 14
   Interrelationships of Discharge, Velocity, Depth, and
       Width in the Orashi and Sombreiro Rivers 15
   Various Combinations of b, f, m Possible and Their Significance 20
   Channel Patterns of the Orashi and Sombreiro River Systems 21
   The Meandering Patterns of the Orashi and Sombreiro
       River Systems 22

3. Areal Aspects of the Integrated Drainage Character of the Ahoada East/West and Ogba Ndoni and Egbema Areas of the Orashi-Sombreiro Plains 27
   The Phenomenon of Flooding in the Orashi-Sombreiro Plains ... 31
   Rain flood (Locally called "Uwhoni" in Ogba) 31
   River Flood, ("miNi-iji, in the Ogba local parlance) or Flood due to the Rise of the Orashi Water Consequent upon the Niger Floods 39
   Flood Probability and Frequency in the Orashi 46

4. Climate Change and Human Factors in the Drainage Characteristics of the Ahoada East and West and Ogba Ndoni Egbema Areas of the Orashi-Sombreiro Plains 51
   Climate Change Factors 51
   Human Factors 53
   Bad drainage and Flooding Effects 58
   Some Examples of Flood Impacts in Study Area (2012) in the words of the victims: 62

5. Suggested Measures for the Improvement of Drainage and Mitigating Flooding Impacts in the Study Area 70
   Flood Risks Management and Mitigation Measures 70
   Summary and Recommendations 75

Bibliography 78

Supplementary Sources 80

Appendices 81
   Appendix A: Fresh Water Discharge Measurements 81
   Appendix B: Computed channel width, meander wave length and amplitude for the orashi and sombreiro 82

Appendix C: Stream Order and Stream Number in the Ahoada Area of the Orashi-Sombreiro Plains + ... 83
Appendix D: Drainage Area and Cumulative channel length in the Orashi-Sombreiro Plains 83
Appendix E: Climatic Data (Rainfall) for Ahoada 84
Appendix F: Granulometrical Composition of Soils** 86
Appendix G: Annual Highest Water Level on the Niger at Onitsha* 87

Synopsis

The Ahoada West and East, and the Ogba-Egbema and Ndoni Local Government Areas (LGAs) within the Orashi-Sombreiro plains are unique in the character of their topography and drainage characteristics. Here, a landscape lacking in significant slope but abundant in scars of flood flows and a network of creeks and rivers is acted upon by an interaction of complex human and bio-climatic factors - rainfall, soils and vegetation - to create a surface drainage character that provides important natural resources but also has become a problem to man.

An objective analysis of the drainage character and the interaction of the hydraulic characteristics and processes of the major drainage channels in the region and impacts on the communities have been attempted in this study. Geologically and geographically, the area set in the Sombreiro-Warri Deltaic plain, is characterized by a strati-graphical sequence marked by alternating sands and clays and an attendant lack of any significant relief.

The consequences of the areal and lineal aspects of the integrated drainage systems of the plain have been analyzed in this study in relation to the total interaction of forces in the area. Furthermore, the influence of man in the interplay of forces is examined and suggestions have been advanced for addressing the acute drainage problems especially flooding of the environment and the communities in the region. The study in these respects is one on the "physical ecology" of the area – the relationship between "water, earth, and man."

Acknowledgement

Many people assisted the author in the study that culminated in the production of this book. Chief among them when he was an undergraduate at the University of Nigeria, Nsukka, Nigeria were Professors G.E. Oformata and late George A. Wigwe. As a student keenly interested in the geology and geomorphology of the upper Niger Delta region, both provided the stimulus that led to my interest in the drainage issues especially periodic flooding of many Ali Ogba communities in the upper Niger Delta region of Nigeria. For their counsel, perceptive criticisms and advice at all stages of my research on the area, I am exceedingly grateful.

Also, I gratefully acknowledge the assistance of the following: Messrs. M.D. Ojuka (Technical Officer) and Eugene J. Kpe (Mapping Assistant) both of the then Surveys Division of the then River's State Ministry of Works, Land and Transportation, Port Harcourt, for making available to me aerial photographs covering the area of study. Messrs J. Blokzijl (Production Geologist) and E.P. Van Maene (Head, Surveys Operations) both of then Shell – BP. Company, Port Harcourt were also very helpful in explaining the geology of the delta environment and in showing me some survey maps and location topographic sheets of some areas covered in the study. And finally, I thankfully acknowledge the help of all others who assisted me with any kind of information and encouragement which I needed during my research on the subject and production of this work.

Introduction

Pleistocene geomorphic processes of the River Niger, under condition of changing sea level initiated a sub-aerial delta environment[1] part of which today constitutes the area defined as Ahoada area of the Orashi-Sombreiro plains in this study (Figure 1). In this region are large and small communities in Ogba Ndoni and Egbema Local Government Area, as well as Ahoada West and East Local Governments. Here, the interplay of past and present fluvial processes of the River Niger and its feeder streams and distributaries have developed a braided channel system. And, factors of reduced gradient, low topographic relief and a rainfall regime marked by all year round rainfall with a double maximum in the summer months coupled with human actions generate within this environment of intricate channels networks, the forces responsible for the spectacular drainage characteristics and at times disastrous flooding that give the area under study its identity.

The purpose of this study is to analyze the drainage of the area in the light of the inter-relationships of the channel systems on the one hand and the general physical environment on the other. Thereafter, it will attempt to investigate other causally related factors including land use patterns responsible for the phenomenal bad drainage topography and flooding of the area as these vary seasonally both in intensity and influence on human lives. Finally, suggestion on possible measures of improving the bad drainage character and mitigation of the impacts of flooding phenomenon of the communities in the area

---

1   J.R. Allen *"Quaternary of the Niger Delta and Its Adjacent Areas: Sedimentary Environment and Lithofacies,"* Bulletin of the American association of Petroleum Geologists, p. 569.

will be proposed.

The Ahoada area of the Orashi-Sombreiro plains covered by this study is taken to mean the area drained by both the Sombreiro and Orashi river systems (Figures 1 and 4) beginning from about the source area of the Sombreiro River, north of the Okwuzi-Aga (Egbema) line and terminating south of Ahoada town along a line traceable through Ochigba and Edeoha. It could be broadly located south of Oguta (Imos State) on the eastern upper flood plain of the Niger Delta where the Sombreiro and the Orashi border it on the east and west respectively, hence the nomenclature: the "Orashi-Sombreiro" plains. It is the area of the former Ahoada Local Government Area (ALGA) which today includes the present Ahoada East and West Local Government Areas, as well as Ogba-Egbema Ndoni Local Government Area (ONELGA).

This area requires the attention of geomorphologists and non-geomorphologists, climatologist and hydrologic engineers and policy makers alike not only for the intrinsic interest which its geomorphic history possesses but because it is the scene of massive oil industry and drainage character whose manifestations have a tremendous bearing on the life of the people who live in the region.

Much of the ideas expressed in this paper have been drawn from direct field observations and by analytical study of aerial photographs which have the merit of portraying the channel geometry of the main rivers in their true perspective. References were also made to some topographical and geologic maps of the area in question. These in combination with the aerial photographs were used for the measurement of some of the quantitative geomorphologic phenomena of the channel systems. Also, the literature produced by NEDECO (1959) "The Waters of the Niger Delta" was the source of some basic data on the hydraulic characteristics of the Niger, Sombreiro and Orashi rivers.

The author, however, regrets his inability to substantiate certain points and observations in this paper with adequate statistical data. This is due to a number of factors, namely: the general dearth of data on the physical and socio-economic phenomena of the area, the destruction of much of what was available in the past during the

Nigerian-Biafra civil war, the impossibility of obtaining adequate quantitative geomorphic analysis of the drainage phenomena of the area because of the seasonality of most of the drainage channels - since these are practically invisible in aerial photographs, due to the masking effects of vegetation which also makes penetration into the bush for on the spot measurements difficult and shortness of the time for the research. But, in response to greater objectivity and to the possible implication for planning and the utilization of rural landscape for effective human occupation, a conscious effort was made to observe, document and measure quantitatively, as far as possible, some of the physical and human processes occurring in the area.

1

The Environmental Background of the Ahoada Area in the Central Orashi-Sombreiro Plains

The Study Area

The region covered in this study falls within Ahoada East and West Local Government Areas, as well as Ogba-Egbema and Ndoni Local Government Area of Rivers State, Nigeria. The three local government areas cover a total of 2,365 square kilometers with a 2016 population estimate of 981,900 according to Nigeria Population Commission. This population represents 5.08% of Nigeria's population estimate for 2016, and 13.4 % of Rivers State 2016 population. The Ahoada areas as defined in this study (Figures 1 and 4) and in terms of its broad environmental setting lies within the catchment area of the Niger/Benue river system (Figure 1). This area, which constitutes the gathering ground of fresh water that drains the plain, has a total drainage area of 730,000 sq. miles. Of this 222,000 sq. miles occur in Nigeria, covering about 60% of the country's total area[2].

---

2   Nedeco, River Studies: Niger and Benue Report, p. 169.

But, in its restricted locale and ecological setting, it is part of the Orashi-Sombreiro plains (Figures 2 and 4). This area spatially and geologically corresponds to the eastern section of the formation commonly known as the "Sombreiro-Warri Deltaic plains"[3] (Map 1). The plain is generally conceived to have originated during a transgressive phase following a major regression during the last Pleistocene glaciation[4] during which " ... the major part of the present Delta was deposited (in this) Wurm I and Wurm II interval."[5] In this context, therefore, the Ahoada area under discussion is a sedimentary environment. And because the older deposits of sedimentary materials now constitute the coastal plains sands bounding the recent Niger Delta area, the Orashi-Sombreiro plains which forms the central portion of the Sombreiro-Warri Deltaic plains occupy what can be considered a transitional belt to the older Niger Delta area and the recent Niger Delta environment.

---

[3] Nedeco, The Waters of the Niger Delta, p. 109.
[4] Idem.
[5] Nedeco, River Studies, Niger and Benue Report, p. 259.

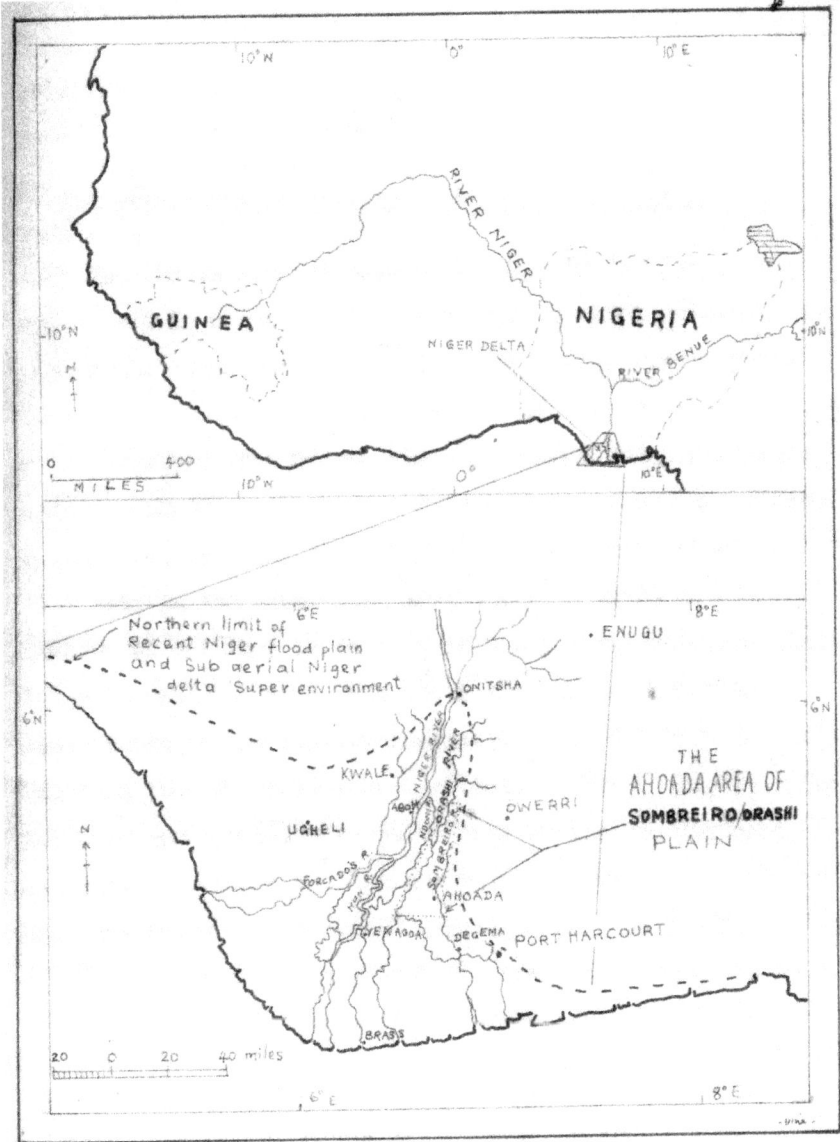

Figure 1 Location of the Ahoada Area of the Sombreiro-Orashi Plain

Figure 2 Map of Rivers State Showing LGAs

Land Forms and Climatic Factors

In the wider context of Nigeria's hydrogeological setting shown in map below, the area covered in this study falls within the country's coastal sedimentary lowlands characterized by low gradient, high permeability and low run off. It also falls within hydrological area V of which Rivers State is a part. Based on the above characterization of the region, the environment consists of three ecological zones, namely:

(i) Orashi and Sombreiro rivers plains;
(ii) lowland forest and farm mosaic, and
(iii) non-tidal fresh water swamp basins which are derived from the geomorphological structures of the region.

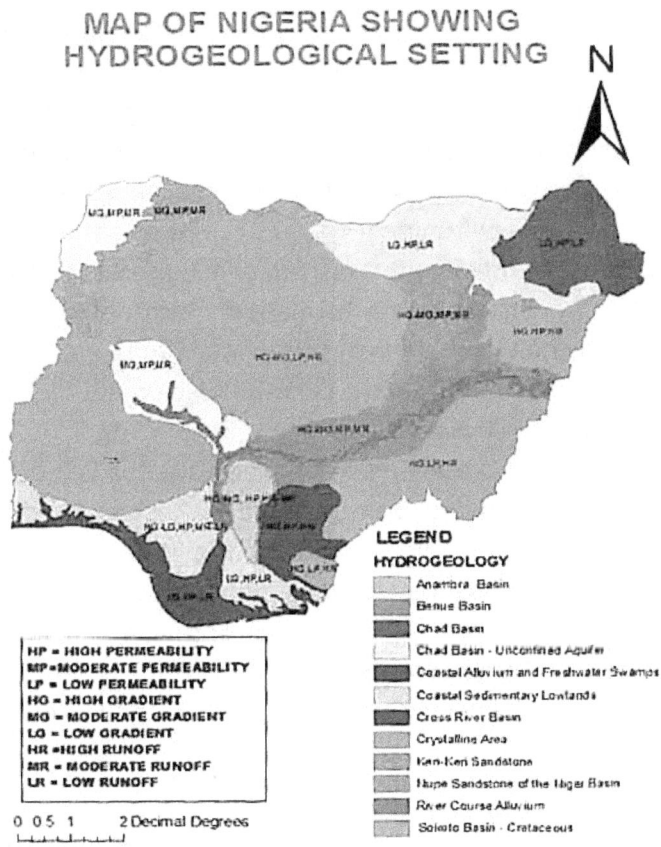

Map 1: Nigeria hydrogeological setting
Source: Nwankwoala, H. O. (2015)

Stratigraphical and Structural Characteristics of the Orashi-Sombreiro Plains

On the basis of the geological setting and mode of evolution of the Orashi-Sombreiro plains [also, the Sombreiro-Warri Deltaic plains,] the Ahoada-Omoku area is underlain by thick deltaic deposits of marine, fluviatile and terrestrial character. These deposits which have an estimated thickness of 5000-8000 meters[6] (16,400ft - 26,240ft) consist of a sequence observed laterally and vertically in the Delta environment. Vertically, they grade from Paleocene marine shale and clays through an intervening unit of alternating sandstones and shale (clays, grit, sands, gritty sands, and sandy clays) otherwise known as Agbada formation to the continental gravelly and sandy terrestrial deposits (the Benin Formation) that constitute much of the surface deposits.[7] This Stratigraphical sequence reflects history of the Niger Delta environment which since its inception is one of a major regression with a gradual off-lap of extensive macro-lenses of sediments.[8] The surface Benin Formation which outcrops over the area is a sandy unit, consisting of continental sands, gravels and clay streaks. The sands at these are cross-bedded, while the clays and sandy clays occur in lenses.

Structurally, the Orashi-Sombreiro plains, like the Sub-aerial Niger Delta "super environment" on which it lies, is characterized by a seaward regional dipping. Discernible tectonic structures are mainly growth faults caused by gravitational sliding and severely faulted roll over anticlines.[9] Within the surface Benin Formation, the upper two-third shows little or no deformation either folding or faulting.[10] This is reflected in the general absence of surface, and to some extent,

---

6   J. Hospers, *"Gravity Field and Structure of the Niger Delta, West Africa"*, The Geological Association of America Bulletin, p. 408.

7   K.C. Short and A.J. Stauble, *"Outline of the Geology of the Niger Delta"*, Bulletin of the Association of American Petroleum Geologists, pp. 634-645.

8   Information Unit, Office of the Governor, Rivers State, *"The Geology of the Niger Delta and Surrounding Areas"*, The Oil Rich Rivers State, pp. 12-13.

9   Idem.

10   Idem.

subsurface structures over the area.

The stratigraphical sequence as previously outlined is generally related to the overall drainage phenomena of the area. The alternation of clays, shale and sandstones provide varying water absorption and retention levels which affect the water table. While the upper sandy units permit easy percolation of water to the pebbly levels which yield ground water, the shale and clayey formations and levels constitute the impervious confining beds. These check infiltration, but promote sub-surface seepage as is common in the swamp basins. Where the relatively impervious shale or clayey sub-stratum occurs near the surface, such area remained permanently under water – a condition that is accentuated during the wet season.

But in terms of the general physiographics of the area, the absence of structure deformations, as well as the lack of sharp lithological and textural contrasts are reflected in the absence of bold relief. So, evident in the area is a monotonously flat and low lying land, honey combed by swamps and a mess of effluents, crisscrossing the plain in varying dimensions and character.

As is shown in figure 3, the highest part of the area is the Omoku-Okwuzi area which attains an elevation of 50ft and above, while the lowest parts lie south. Ahoada town, itself in the south-east, is 30ft above sea level.[11] The Ubeta-Oyiba section, south-west of Ahoada town ranges between 24ft and 26ft above sea level.[12] Computed average gradient of the area taken in a north to south direction from Omoku to Ahoada (a distance of twenty miles) gives a value of (or 1 foot rise in 1 mile). Thusly, this illustrates the near level topography of the area, in an area of low amplitude of relief and consequently minimal, if not absent slope. These topographic characteristics enhance the intricate channels system of the area which in combination with some of its climatic factors (soil, vegetation and rainfall) influence tremendously the nature of drainage in the area. This influence

---

11 This was computed from Map of Heights in "*Gravity Field and Structure of the Niger Delta*" by J. Hospers.
12 Computed from Location Topographical Sheet, Ahoada North Location by S-BPC Topographical Dept. Port Harcourt.

is achieved by the way they facilitate or hinder run off, infiltration and or stagnation of water in the low lying areas. They are also functional to the intricate mechanisms of the major drainage channels and their patterns in space as the subsequent section of the essay would show.

The Major Drainage Systems in the Ahoada East and West and Ogba-Egbema Areas of Orashi-Sombreiro Plains

Although the surface morphology of the central Sombreiro-Orashi plains of the Ahoada Area is characterized by abandoned water courses, meander loops, swamps and creeks, the major source of fresh water over the areas remains the Orashi and the Sombreiro river systems (Figures 3 and 4). The former maintains an intimate relationship with the Niger River system. In order to appreciate the fluvial processes of these rivers, and to acquire a better understanding of the overall drainage character of the area, a consideration of the primary river systems in terms of their hydraulic mechanisms in the area is pertinent.

The Environmental Background of the Ahoada Area in the Central Orashi–Sombreiro Plains

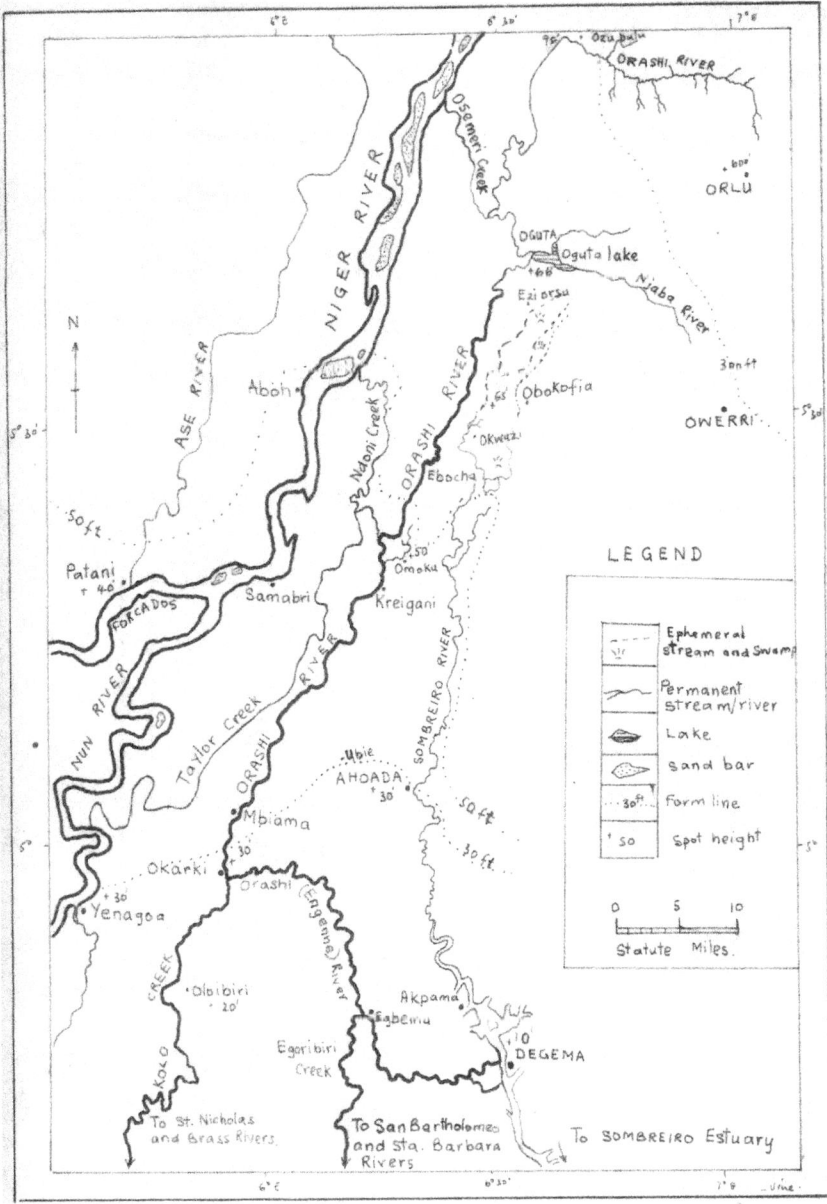

Figure 3 Map of Heights and Part of the Niger Orashi and Sombreiro River Systems

Drainage Characteristics and Flooding Problems

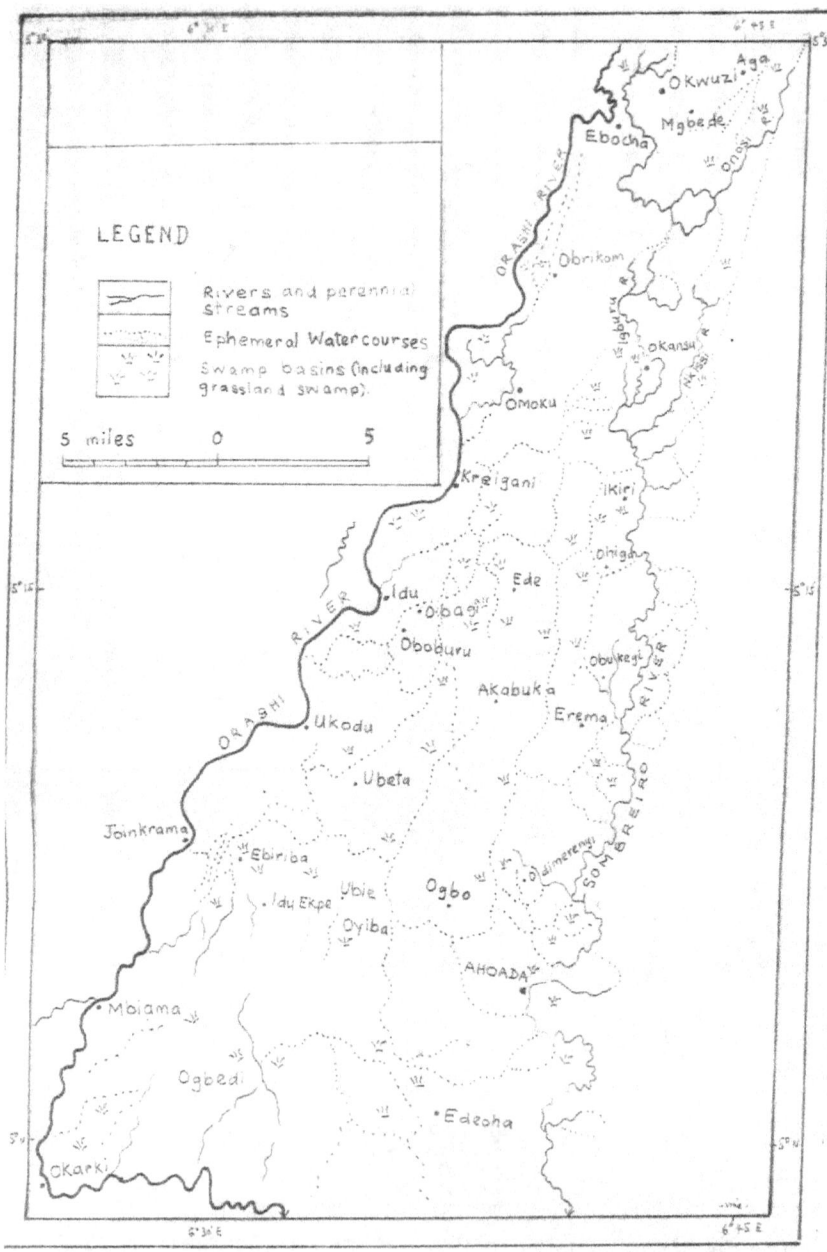

Figure 4 Drainage in the Ahoada Area of the Sombreiro - Orashi Plain

The Environmental Background of the Ahoada Area in the Central Orashi–Sombreiro Plains

The Orashi River System

The Orashi river rises as a deep gorge-like stream from the neighborhood of Dikenafai, near Orlu (Imo State) (Figure 3). Its headstreams on the south-west flanks of the Awka-Orlu uplands constitute many gully-like streams cut deep into the Bende-Ameki and the Agulu Nanka sandy formations. On reaching Ozubulu near Nnewi in Anambra State, the Orashi enters the extensive alluvium veneered Niger flood plain where it makes a dramatic turn southwards, flowing parallel to the Niger rather sinuously on the alluvium filled plain. This mode of flow derives from a number of factors. Firstly, the Orashi having descended from the highland (height of land around Ozubulu being about 100 ft above sea level) onto the low lying Niger River plain (about 50-75 ft above sea level) experiences an abrupt change of slope and so seeks characteristically 'an understanding' with this slope. Here, it cuts it's a path at the zone of contact between the coastal plains sands formation and the flood plain. Secondly, the Niger floods continuously build up the immediate adjacent lands and thereby make these areas higher than the 'backlands' (areas forming the back swamps beyond the levees). Consequently, the Orashi does not flow directly to join the Niger at an angle but meanders as a mature stream (maturity having been imposed on it by the changed topography parallel to the Niger River in consonance to the 'law of delayed junction') (See Figure 3).

North of Oguta Lake, the Orashi is joined onto the Niger River by the Osemiri stream. Flowing by the blue Oguta Lake, the Orashi is later connected with the River Niger by an interjacent stream, the Ndoni (Onita) creek through which the River Niger water enters the Orashi in the central Orashi (Omoku to Ahoada area). Below Oguta and in particular the area from Omoku, through much of the western part of Ali Ogba (Kreignai, Idu, Obigbor, Ohali Elu and Ohali Usomini, Oboburu, Obagi, Ede, Obite), Ndoni and Egbema communities, the Orashi meanders southwards through its own flood plains (now superimposed on the Niger River floodplain). Towards Okarki, (the limit of the area defined as central Orashi-Sombreiro plains covered in this study) some of its water flows south as the Kolo Creek, while the main river (now Engenni River) turns eastwards passing through Ikodu, and later southwards to the neighborhood of Ighom near Egbema, where it turns eastward again to join the Sombreire below Degema. All through the sixty-seven-mile length of its course in the Ahoada area, the Orashi flows more or less parallel to the River Niger and depending on both local rains and the River Niger for its discharges.

The Sombreiro River System

The Sombreiro river system forms the eastern limit of the area under study and the second major river that drains it. The source area of the Sombreiro is traceable to east of the Ezi-Orsu-Oborotu axis south of Oguta (Figure 3). Here, below the 50ft, contour of the sandy highland, there occur swamp basins from which issue the two tiny head streams of the river locally known as the Ujiji and the Onosi. These streams dry up to mere stagnant pools in the dry season, while maintaining a high water level in the rainy season. Only at the neighborhood of Oborotu and Obokofia (Egbema area) southwards in respect of the Ujiji and Onosi, respectively that permanent water flow is maintained in these channels all the year round.

The Ujiji flowing from east to the southwest is separated from the swamp and alluvium-fringed valley of the Onosi by a relatively high and narrow sandy land with an average elevation of about 65ft above sea level in the Ukwugba area (Figure 3). These two streams meet south of the Mgbede-Aga locality in another extensive swamp basin. It is out of this swamp basin that the headstream of the actual and present day Sombreiro river begins as is often illustrated in many topographical maps of the area. The two prominent headstreams of the Sombreiro identifiable from the name of interlacing creeks at the southern extension of the Mgbede-Aga swamp basin are the Igburu and the Nkissi. These are connected with the earlier mentioned Ujiji and Onosi systems, respectively. Here, drainage is confusingly complex because of the crisscross of the creeks and the thick vegetal cover which makes penetration into the area and the identification of the drainage channels in aerial photographs difficult.

From the interconnected network of creeks and swamps south of the Ngbede area, the Igburu and the Onosi/Nkissi systems join north of Ikiri (Figure 4) to flow rather tortuously as the upper Sombreiro. This mode of flow results from the responses of the hydraulic characteristics of the channels on the structure of the area. At this headstream section of the Sombreiro, the river is narrow and shallow, being only 10-15 ft. deep. But southwards, in the neighborhood of Ahoada, it is about 20ft deep[13]. Its total length in this area of study is about 39 miles.

After the confluences of the Igburu and Nkissi/Onosi systems, the Sombreiro proceeds in a southerly direction as a small meandering river, receiving a few short side creeks draining the adjacent swamps. This it does until it reaches the recent Niger Delta area where it debouches into a lake near Akpama (Figure 3). On leaving the lake, the river is drained by a number of tidal creeks before it finally enters the ocean through the "Sombreiro Entrance."

---

13   Nedeco, The Waters of the Niger Delta, p. 202.

2

Some Aspect of Hydraulic Geometry of the Main Drainage Systems of the Studyarea: Orashi and Sombreiro Rivers

WHAT LEOPOLD AND Maddock (1953) termed hydraulic geometry is the graphical analysis of the hydraulic characteristics of a stream channel. It explains the quantitative relationships among channel widths, mean depth, mean velocity and suspended load as they vary with discharge both at a given stream cross section (referred to as "At-a-station measurements"),[14] as well as at different cross sections along the stream length (otherwise known as "measurements in a downstream direction").[15] These hydraulic characteristics of river channels vary with discharge as simple power functions. Discharge is regarded as an independent variable, while the other series: width, depth, velocity and load in their interrelationships with discharge are taken as dependent variables.

Many terrestrial streams develop and propagate consequent upon their intrinsic hydraulic conditions. By this, they create their own geometric

---

14 L.B. Leopold et al., Fluvial Processes in Geomorphology, p. 216.
15 Ibid., p. 214.

properties.[16] So, comparisons of various cross section characteristics along the length of a stream by a detail of the variations of its hydraulic characteristics in a down-stream direction illuminate the current hydrological and fluvial processes of the river channel. Furthermore, it is the dominant discharge (equivalent to flow at bankfull stage) that determines and maintains the form of a river channel. To apply these principles to a description of the hydraulic characteristics of the Sombreiro and Orashi requires considering variations of the hydraulic geometry of the Sombreiro and Orashi at Ahoada and Okarki, respectively.

Since data on fresh water measurements exist only for the Orashi at Okarki and the Sombreiro at Ahoada, for this area of study, it is impossible to include an analysis of the downstream hydraulic characteristics of the channels in this exercise. So, only "At-a-Station Measurements" would be used for the hydraulic geometry analysis of the main drainage channels.

Interrelationships of Discharge, Velocity, Depth, and Width in the Orashi and Sombreiro Rivers

Figure 5 represents data which show changes in width, depth and velocity for the Orashi at Okarki and the Sombreiro at Ahoada. The variables related to discharge as simple power functions are:

$W = a Q^b$ ... (i) b, is the rate of change of width with respect to discharge
$D = c Q^f$ ... (ii) f, is the rate of change of depth with respect to discharge
$V = k Q^m$ ... (iii) m, is the rate of change of velocity with respect to discharge.

Where Q is discharge, W, D, and V represent the water surface width, mean depth and mean velocity respectively.[17]

The coefficient a, c, and k are constants and b, f, and m are exponents representative of the channel cross section[18]. Data for the

---
16 R. Pestrong, Development of Drainage Patterns on Tidal Marshes, p. 42.
17 Leopold et al., op. cit., p. 215.
18 Idem.

computation (see Appendix A) were obtained from measurements made by NEDECO (1959) at gauging stations where fresh water discharges had been measured at regular intervals.[19]

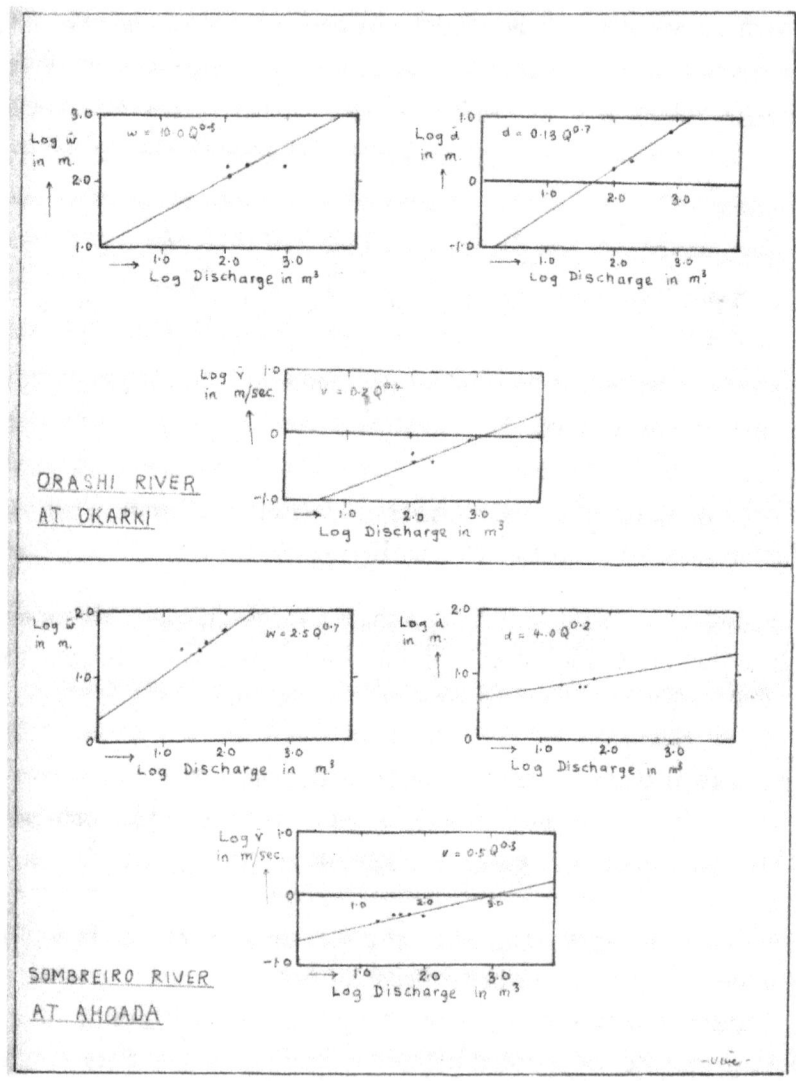

Figure 5 Changes in depth, width, depth and velocity with discharge at -a- station

---

19 NEDECO, *"Freshwater Discharge Measurements"*, Waters of the Niger Delta, p. 42.

The graphs show a general tendency for width, depth and velocity to increase as discharge increases. The regression equations obtained by the method of least squares from the repetitive data for the two river systems are tabulated as follows.

TABLE 1 HYDRAULIC EQUATIONS FOR:

ORASHI RIVER	SOMBREIRO RIVER
$W = 10.0\ Q^{0.5}$	$W = 2.5\ Q^{0.7}$
$D = 0.13\ Q^{0.7}$	$D = 4.0\ Q^{0.2}$
$V = 0.2\ Q^{0.4}$	$V = 0.5\ Q^{0.3}$

TABLE 1: EXPONENTS AND CONSTANTS FOR THE EQUATIONS DERIVED FROM GRAPHS IN FIG. 5

EXPONENTS	ORASHI RIVER	SOMBREIRO RIVER	CONSTANTS	ORASHI RIVER	SOMBREIRO RIVER
b	0.5	0.7	a	10.0	2.5
f	0.7	0.2	c	0.13	4.0
m	0.4	0.3	k	0.2	0.5

The values of the exponents as determined in the study characterize the geometry of the channels in relation to the changing discharges, and the resistance to erosion associated with the character of the bed and banks.[20]

For the Orashi River, changes in width, depth, and velocity with increasing discharge are varied and minute. The exponents (Table 1) reveal for any value of discharge a relatively faster change in depth, than in width and velocity. In other words, there is increase in either width of channel or its velocity, as the river increases its discharge. The meager depth increase is attributable to the scouring action of bed materials and the consequent deepening of the channel depending on the resistance of the bed and bank to corrosion. This scouring action involves a loss of much stream energy for discharge, hence the

---
20 Leopold et al., op. cit., p. 217.

low exponent relating velocity with discharge. It is this phenomenon – relatively higher rate of change in depth than rate of change in width and velocity with increasing discharge – which accounts for the tendency for the Orashi to develop a channel characterized by steep sides (box-like channel) in some localities such as Idu and Ukodu. In these areas, the eastern banks of the channels have sandy clays which form cohesive materials facilitating the development of the "box-like" type of channel (Figure 6: top section).

The values obtained for the constant (Table 1) indicate the ratio between the changes in width, depth, and velocity, and the increase in discharge. The ratio of width to discharge is the highest of all, constant "a" being 10.0. This means that at any instant, the rate of change in width to discharge raise to the power 0.5, is 10.0 to 1. But that of depth to discharge is 0.13 to 1. Therefore, any slight increase in discharge will result in a faster change in width than in depth in the neighborhood of Okarki. This analysis reflects the potentiality for lateral outflow of the Orashi in this section of the river with increasing discharge.

In terms of the velocity, the proportion of velocity change to discharge is very small, being 0.2 to 1. This implies that for any appreciable change in the velocity to occur, there must be a high increase in the discharge.

Exponential values computed for the Sombreiro differ slightly from those of the Orashi. But values obtained for the constants are remarkably different for the two river systems. The exponential figures for the Sombreiro show that width increases faster than depth and velocity for any unit of discharge. The difference between changes in depth and velocity is very slight, being only about 0.1. This relatively rapid change in width as against changes in depth with increasing discharge has a number of implications.

First, the high value of change in width and the low value for change in depth indicate a tendency for the channel to widen and become shallow. Furthermore, the rapid rate of increase in width gives an idea of the general shape of the Sombreiro channel of that of a "wide dish" (Fig. 6b). So, the Sombreiro has a potentiality for easy lateral outflow if the discharge increases significantly. On the other hand, the low rate of change in velocity for a given value of discharge indicates slow flowage, minimal energy and absence of bank caving.

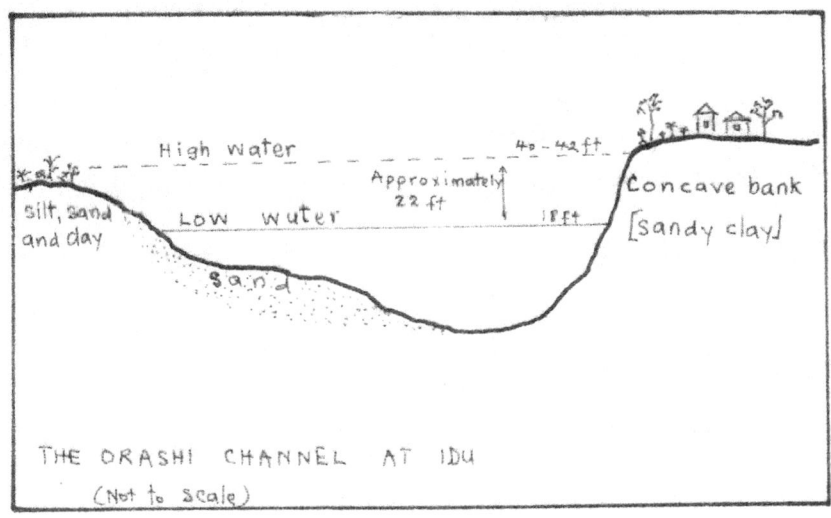

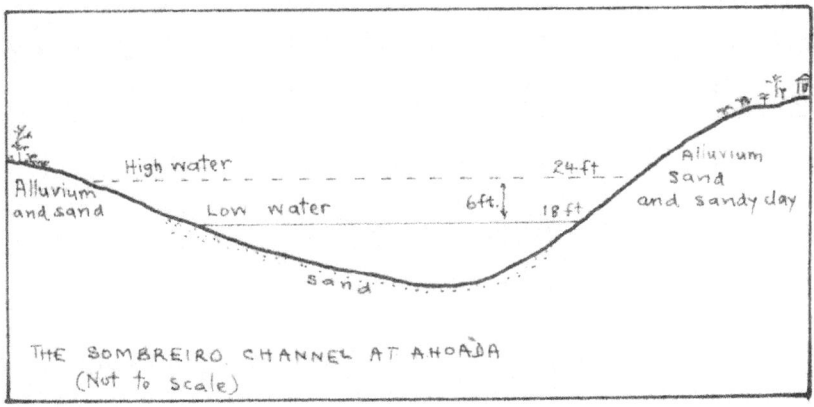

* Heights estimated to mean sea level.

Figure 6 Channel forms of the Sombreiro and Orashi Rivers

Constants a, c, and k for the Sombreiro have values of 25, 4.0, and 0.5. The highest value is for depth which is directly proportional

Drainage Characteristics and Flooding Problems

to discharge raised to the power of 0.2 – the constant of proportionality being 4.0. In other words, at any instant, the ratio of depth to discharge raised to 0.2 in the Sombreiro is 4.0:1. This further means that the Sombreiro could increase in depth significantly with a unit increase in discharge. This reflects the narrowness and shallowness of the river in this area. While the ratio of width to discharge raise to the power of 0.7 is 2.5 to 1, that of velocity to discharge raised power of 0.3 is 0.5 to 1. The importance of these figures lies in the fact that width and velocity, just like depth, change faster than discharge.

The m/f ratio i.e. $\dfrac{\text{rate of increase of velocity with discharge}}{\text{rate of increase of depth with discharge}}$

Computed for the Orashi and the Sombreiro gives the following values:
Orashi   0.6
Sombreiro 1.5.

These figures have relevance to sediment load which is related to flooding. Because the values are low generally, they indicate that sediment load increases less rapidly with increasing discharge. So, with this characteristic low sediment load, the channels are unable to effect much bank erosion either laterally or vertically as reflected in the relationships between discharge and channel width, depth, and velocity discussed previously. Consonant to the low sediment load of the channels, which also implies low velocity; the channels are liable to overflow their banks in periods of high discharges.

Various Combinations of b, f, m Possible and Their Significance

For the Orashi river, a possible b, f, m combination with values for b=0.4, f=0.5, m=0.7 would indicate a faster increase in channel depth than in width with an increasing discharge. The high value of m (0.7) shows that the velocity increases fastest of all the parameters. In such an instance, the channel form will tend to be box-like and there would probably be a slightly higher increase in the sediment load since m/f ratio of 1.4 might be obtained. Only in cases of very high discharges does flooding affect the adjacent lands. But if (for the Orashi) the combination of the b, f, m is thus: b = 0.7; f = 0.4; m = 0.5, the channel will tend to widen at a faster rate than it is deepened with increasing discharge. The low f value and a high b value combining with an m value of 0.5 would promote disastrous flooding with increase in discharge.

In the case of the Sombreiro river, a combination of b, f, m values: b=0.3; f=0.7; m=0.2 would indicate a higher rate of increase in depth than in width and velocity as the discharge increases. Under this circumstance, the Sombreiro channel would change its form from being wide dish shaped to the entrenched type. The low velocity would result from energy being expended by the river in deepening its channel. Thus, overtopping of the banks would be a rare occurrence. But if the combination changes to b=0.2, f=0.3, m=0.7, then velocity would increase fastest than both width and depth with increasing discharge. And, with a possible m/f ratio of about 2.5 sediment load increase would be high as the discharge increases. This would probably lead to bank caving, depending on the resistance of the bed and bank materials. But with such a high velocity, nearly all the materials of the channel would be moved at all stages of the discharge.

Channel Patterns of the Orashi and Sombreiro River Systems

Channel patterns mean the "configuration of a river as it would appear from an airplane"[21]. The most common channel patterns identifiable on the Orashi and Sombreiro river systems are the braided and meandering patterns.

Downstream of Ukodu on the Orashi, there is an appreciable amount of channel materials which give rise to braiding. This phenomenon results from selective deposition of the channel materials

---

21   Ibid., p. 261.

(sediment) and the formation of central sandbars resulting in a diversion of flow with increasing erosion attack indifferently on the channel banks.

This accounts for the sandbars between Ukodu and Okarki. In these localities, the sand bars form good fishing grounds, being submerged during high water stages but remain exposed to view at low water. Statistical support for the increase in bed load and sand transport on the Orashi is derived from the data for the gauge below Egbema. In 1959 and 1960, bed load transport on the Orashi amounted to 12,500m^3. (260, 00 cu ft) for the station. A yearly suspended load to the tune of about 7,500m^3 (about 260,000 cu ft)[22] has also been measured.

For lack of data, it is impossible to statistically show bed load transport for the Sombreiro as a basis for discussing the braiding phenomenon of the river. Nevertheless, observations made during the study reveal that the Sombreiro is characterized by anastomosing channels. Between these, there are small islands which are thickly forested and rarely covered by water except during very high water stages. Very good examples of these occur at Orieke near Ahoada, at Ahoada town area itself and in the Okansu-Uju-Agbada area near Omoku.

The Meandering Patterns of the Orashi and Sombreiro River Systems

A braided channel invariably may show curves with a characteristic radius to channel width relationship that identifies some reaches as meandering. This shows that there is some relationship between braiding and meandering in rivers. Because a braided reach is wide and shallow, and the channel banks unstable, the rate of sediment transport may then be relatively low. So, deposition becomes characteristic of such a reach. And, if by this, alternating "hollows" and "shallows" appear on the stream bed to control the water flow direction, the stream begins to swing from side to side; in other words, meandering ensues.

---

22 Nedeco, Waters of the Niger Delta, p. 105.

But for a river to be described as meandering its "ratio of channel length to down valley distance"[23] should be as follows: Sinuosity ratio has been judged to be equal to or exceed 1.5. Sinuosity ratio varies in rivers from unity to 4 or more and when S., r. is greater than 1.5 the river is meandering, but when S.r. is less than 1.5 it is merely sinuous or straight.[24]

TABLE 2:

River	Range of Sinuosity Ratio (Sr)	Average Sr
Sombreiro	1.5 – 4.0	2.4
Orashi	1.5 – 2.5	1.7

Table 2 above shows computed sinuosity ratio for various sections of the Sombreiro and Orashi channels. It also reveals that the average sinuosity ratio of the channels sections exceeds 1.5, thus characterizing the two channels as measuring, at least in consonance to the varying values for b, m. f and a, c, k discussed and the observed braided pattern of the channels.

Beside sinuosity, the attributes of a meandering channel include the nature of amplitude, wave length, and radius (illustrated in Fig. 7). Of all of these, the meander length (wave length) is the most significant being fundamentally related to bed width and discharge.[25]

The empirical relationship between meander wave length and channel width on the one hand, and amplitude to channel width on the other hand has been graphically illustrated in Fig. 8 for the Orashi and Sombreiro.

As Fig. 8a shows, wave length decreases with increasing channel width for both the Orashi and the Sombreiro. The regression equations obtained from the graphs are presented in Table 3 below.

---

23 Leopold et al., op. cit., p. 281.
24 Idem.
25 Ibid., p. 296.

TABLE 3: EMPIRICAL RELATIONS BETWEEN SIZE PARAMETERS FOR MEANDERS IN THE ORASHI-SOMBREIRO RIVERS
(See also Appendix B)

RIVER	MEANDER LENGH TO CHANNEL WIDTH ($^mL:W_E$)	AMPLITUDE TO CHANNEL WIDTH ($A:W_E$)
Sombreiro	$m_L = \log 5.6\, w^{1.3}$	$A = \log 0.8\, w^{0.7}$
Orashi	$m_L = \log 5.9\, w^{0.7}$	$A = \log 2.9 w^{0.3}$

The exponents in the equations are varied for the two channels in terms of their wave length and channel width relationships. For the Orashi, this is 0.7; while for the Sombreiro, the value is 1.3. This indicates a fuller development of meanders is the Sombreiro than in Orashi. But the decreasing wave lengths vis-à-vis increase in channel width for both rivers also implies that meandering is progressive in both rivers.

The meanders tend to develop large loops with high amplitudes as the channel width increases consonant to the discharge, width, depth, and velocity relationships earlier discussed. This is illustrated by the graph of amplitude to channel width (Fig. 8b). Although the exponential figures vary from 0.7 for the Sombreiro to 0.3 for the Orashi (Table 3) the increase in amplitude with channel width testify to the fact that the meanders are fully developed. Besides, it points out that flooding of these channels with increasing discharge is of high probability, since the channels tend to widen as the wave length decrease. The generally higher exponential values for the Sombreiro in the two equations reveal that the size parameters of the Sombreiro meanders are greater than those of the Orashi. This probably results from the influence of the vestige of the ancient Sombreiro meanders still found in the valley today on the observed meander patterns.

The occurrence of full meanders, cuts off channels and anastomosing channels in the Sombreiro and Orashi valleys when considered in relation to the hydraulic characteristics (relationships among channel width, depth, velocity, and discharge) already discussed seems to indicate that these meanders were initiated at a period of higher discharges than at present. This then lends support to the idea that the Orashi is gradually becoming obsolete because it no more discharges much of the River Niger water as it did about a century ago.[26] It is conjectured that the continuous raising of the River Niger banks is accountable for the diminishing importance of the Orashi in the discharge of the Niger water. But a probable result of these phenomena may be climatic change rather than any kind of capture or structural readjustment.

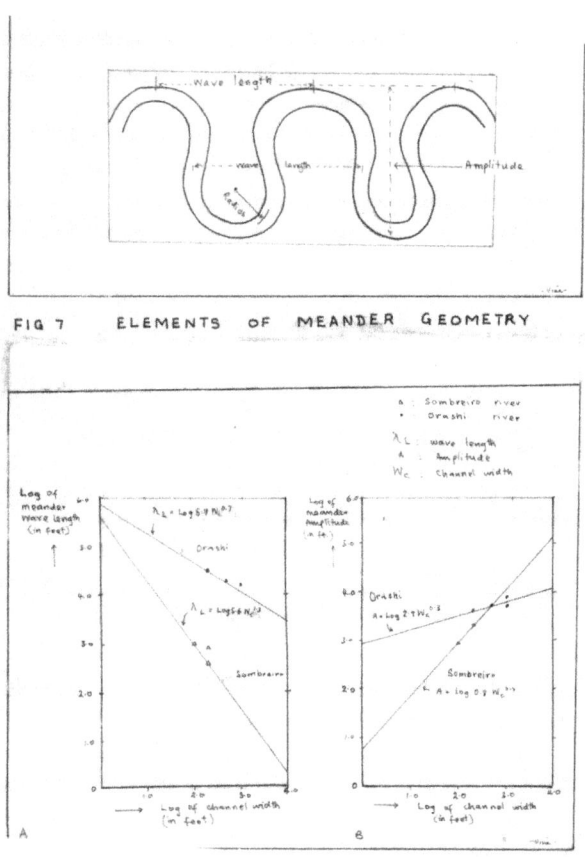

Figure 7 (Top) Elements of Meander Geometry and Figure 8 (below) Relation of Meander wave length and amplitude to channel width of the Orashi and Sombreiro Rivers

---

26 Nedeco, Waters of the Niger Delta, p. 109.

These observations' similarities validate the widely held notion that the Sombreiro lost its main water when it is was cut from the Niger River system in contemporary recent times.[27] It testifies too that the present Sombreiro occupies a valley not cut by it. On this basis, therefore, the Sombreiro can be considered an "under fit" stream meandering within a valley meander and apparently too small for the valley in which it flows today. Thus the considerably low discharges of the present Orashi and Sombreiro have made them less active in initiating new meanders.

Contrarily, what have become evident are relics of earlier meanders preserved beyond the present channels. These abandoned meanders have high water levels in the rainy season because they easily become inundated. Otherwise, they are densely covered by forest and remain marshy almost all the year round.

---

27   Ibid., p. 108.

3
Areal Aspects of the Integrated Drainage Character of the Ahoada East/West and Ogba Ndoni and Egbema Areas of the Orashi-Sombreiro Plains

THE ORASHI-SOMBREIRO PLAINS comprising Ahoada West and East constitutes an "open" basin area. It is therefore, difficult to study the drainage network of the area especially in the absence of good up-to-date topographical maps. In spite of the problem of categorizing the respective stream orders in the area, samples taken and represented in Figure 9 (see Appendix C) obey the law of stream number (Horton, 1945).[28] From the graph, an average bifurcation ratio (ratio of streams of any given order to the next number in the next lower order) can be read. This represents the slope of the line relating number of streams to streams order.[29] Here, its value is 3.2. This is comparable to values obtained in other environments such as the San Francisco Bay (USA) where drainage networks have

---

28 Leopold et al., op. cit., pp. 137-138.
29 Idem.

Drainage Characteristics and Flooding Problems

been studied by Pestrong[30]. It also falls within the range of values (between 3 and 5) characteristic of natural streams.[31]

In terms of the relationship between stream length and basin area, the graph in Fig. 10 (see also Appendix D) defines a line whose equation is:

$$L = 1.3 \, A_d^{0.6}$$

Where L is cumulative channel length in miles and A_d the drainage area in square miles. The "constant of channel mainstream" implicit in this parameter is an important determinant of the area necessary to maintain a unit of drainage channel. From the graph, it can be read that on the average, a drainage basin of 1 square mile would contain about 1.32 miles in length of channel. This compares favorably with that described by Hack (1957) in the Shenandoah Valley (USA).

---

30   Pestrong, op. cit., pp. 36-37.
31   A.N. Strahler, Physical Geography, (2nd Ed.), p. 377.

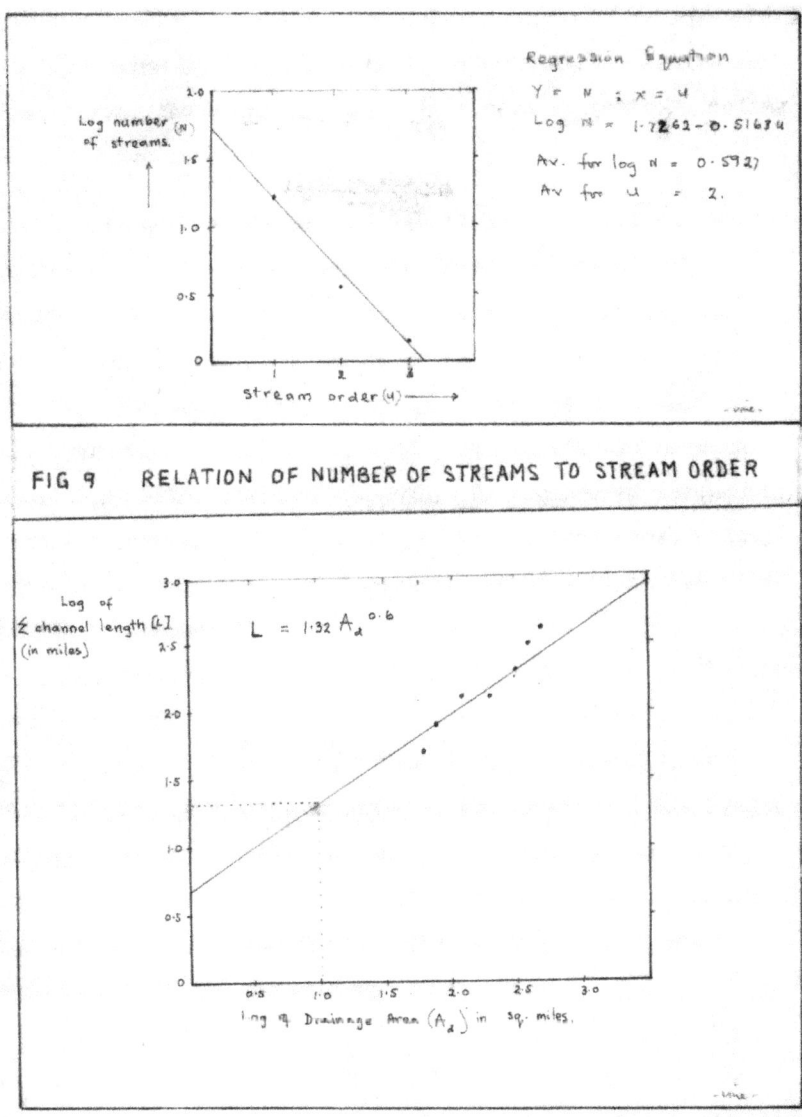

Figure 9 Relation of number of streams to stream order and
Figure 10 (bottom graph) Relation of channel length to drainage area

where drainage basin of 1 square mile would contain 1.4 miles in length of channel.[32]

---

32   Leopold, et. al. op.cit., p.145

Drainage Characteristics and Flooding Problems

The exponent 0.6 indicates that as the size of the area increases, the drainage basin elongates, while the length of the channel increases faster than width. Generally, the elongation of the basin area is at the expense of the length of the lower order streams – in this area it is the first order streams which are many but short in length. Consequently, the lengthening stream (master stream) tends to meander consonant with the elongating basin area.

The concept of drainage density as defined by Horton (1945, p. 283) and meaning "the rate of the sum of channel length in miles to the total basin area in square miles" further illuminates the drainage character of the Orashi–Sombreiro plains of the Ahoada area as defined in this study. With a drainage area of about 400 square miles and a cumulative channel area of about 400 square miles and a cumulative channel length of about 186 miles, this area possesses drainage density which approximates to 0.4649 (0.5). Since most of the ephemeral streams tend to dry up in the dry season (December to February/March) this period is characterized by a drainage density of about 0.4. The drainage density value of 0.5 means that for each square mile of land, there is evident in the landscape a length of half a mile of drainage channels which would be filled with water in the rainy season. The low drainage density figure, probably, is due to the sandy land surface which ordinarily promotes rapid infiltration of water.

Drainage density affects the run off pattern in an area. A high drainage density removes surface run off rapidly decreasing the time difference between rainfall and the resulting stream run off and also increasing the peak of the hydrograph.[33] Similarity, the relationship of total stream length of all orders to basin area provides parameters which control the texture of landscape dissection and spacing of streams thereby giving an idea of run off pattern in an area. Moreover, the bifurcation ratio (earlier discussed) influences the landscape morphometry, as well as controls the peakedness

---

33 R. J. Chorley, (Ed.): Water, Earth and Man, p.81

of the run off hydrograph.[34] With a high bifurcation ratio, run off hydrograph flattens and vice-verse.

The interplay of these features – low drainage density vis-à-vis the bifurcation ratio (3.2) and the characteristic average length of channel per square mile of area (providing the constant of channel maintenance) as discussed imply some degree of inefficiency in run off and a high water infiltration in time and space. The features are casually related to the bioclimatic factors of the environment in effecting the evident high degree of inundation of depressions and low-lands and soil saturation which give character to the landscape of the Ahoada Ogba-Egbema Ndoni area as would be shown subsequently.

The Phenomenon of Flooding in the Orashi-Sombreiro Plains

Flooding is a natural process which results in submergence of land under water especially due to heavy rainfall or river overflowing. In many cases flooding as an overbank flow of drainage channels, is a natural characteristic of rivers which receive more water than can be carried within their banks. In the study area, flooding is extensive. Annually, two distinct types of floods occur varying remarkably in factors causing them, as well as in their periodicity. The two types are rain flood and river flood. The former depends on the relationship between rate, duration and extent of rainfall and the condition of the land in which the rain falls and accumulates. The latter depends on the discharge and channel characteristics of the Orashi in relation to the discharge of the River Niger system.

Rain flood (Locally called "Uwhoni" in Ogba)

As its name implies, rain flood is the inundation of the land due to local rainfall. It is intrinsically related to the rainfall characteristics

---
34   Idem, p.81

of the area and the conditions of the land-geology, physical relief, soil, vegetation and the texture of landscape dissection by drainage channels. So, a consideration of these factors in their interrelationships would facilitate the appreciation of the nature of rain flood in this area.

Climatically, the Ahoada East and West – Ogba Ndoni Egbema region is characterized by a mean annual rainfall that varies from 80" to 100" northwards to about 200" – 120" southwards (Fig. 11). Also, illustrated in Fig. 12 are the monthly graphs of rainfall totals, number of rainy day and rainfall intensity for a year. From the graphs, the intensity of rainfall varies inversely with the number of rainy days and the average monthly rainfall. While the latter two increase, the former decreased, showing a negative correlation.

TABLE 4: RAINFALL CONSTERNATION IN AHOADA

Total yearly rainfall: approximately 101.1 inches
Rainfall concentration, May – October: approximately 79.8 inches
May – October rainfall as percentage   Of yearly rainfall = 67% (see also Appendix E)

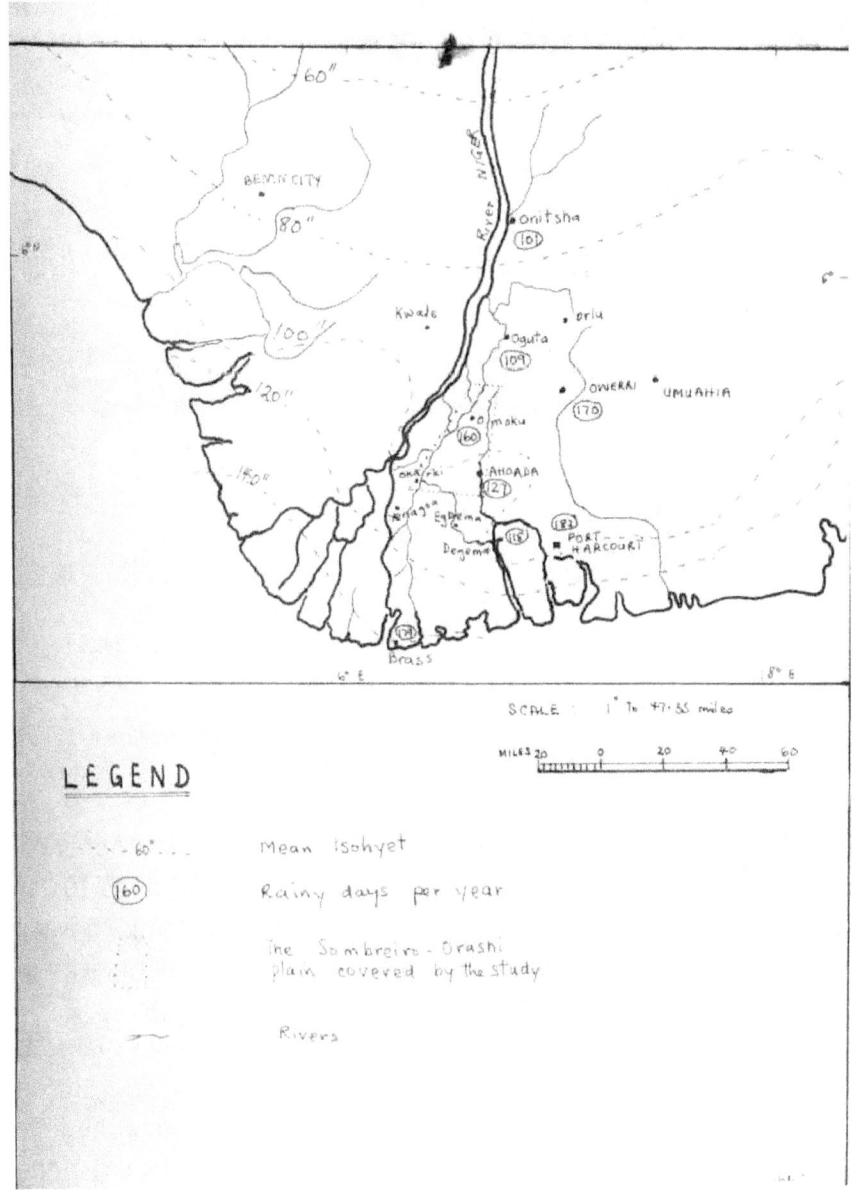

Figure 11 Rainfall Isohyets and number of rainy days

Drainage Characteristics and Flooding Problems

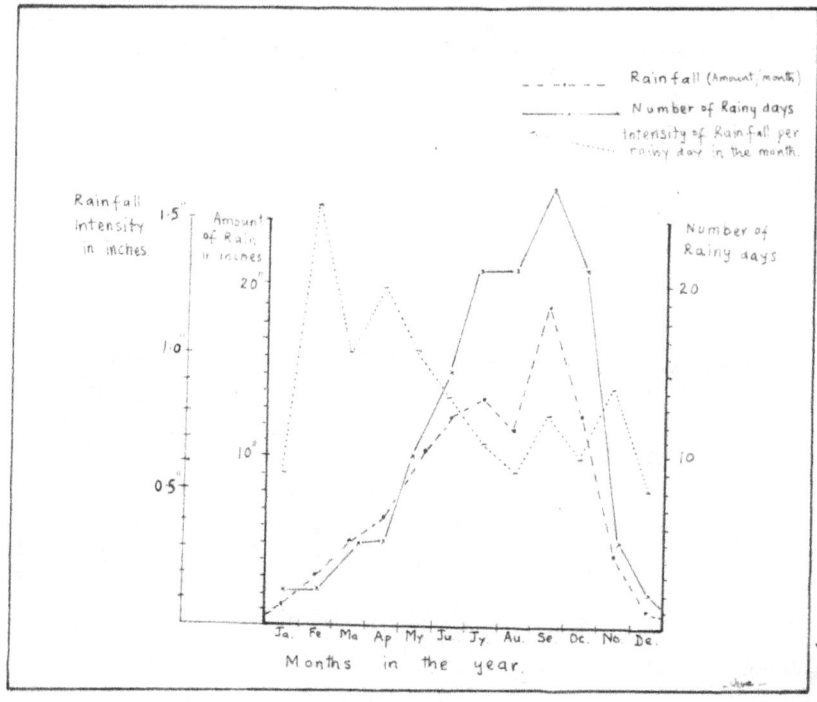

Figure 12 Mean monthly rainfall, number of rainy days and rainfall intensity for Ahoada

Ahoada town has an annual rainfall variability of 11.3% (Appendix E); from Table 4 above, about 79.8" (equivalent to 67%) of its 101.1" annual rainfall is restricted to the period May-October. This shows a disproportionate distribution which is reflected in the periodicity of the drainage problems of the area as would be shown later.

Inadequate soil samples surveys of the Ahoada area makes it difficult to give a detailed analysis of the varying soil types. But on the basis of personal observations of samples taken during the field study and strengthened by analysis made by experts on the soil of the southern part of the Sombreiro-Warri Deltaic plains, the soils of this area belong to the broad ferrallitic group. With such-types as hydromorphic or vertisols on depressions and flooded patches of the land, the soils generally show close relationships with the variations in the topographic character of the plain, as well as with the parent materials which give evidence to its geology.

Figure 13 shows cumulative frequency graphs of soil samples taken in two environments: a well-drained site and a badly drained lowland area. The graphs reveal a normal distribution of soil particles with very high proportion of fine sand in the two environments. A comparative study of the two environments by reference to the graphs in Fig. 11 and data in Appendix F reveals that the sand fraction exceeds 50%. While fine sand exceeds coarse sand both tend to decrease with depth while the clay content increases from the surface downwards. The soil characteristics of these environments bear out relative impermeability and so selective flooding of the area in time and space.

Variations in topography and soil character are reflected in the vegetation observed in the area of study. Much of the area is covered by rain forest with a high regenerating capacity. The drainage channels and swamp basins are covered by thick jungles. While most of the swampy environment have high forests (Picture 1) some have grassy vegetation. These grassland swamps are localized in their occurrence. But they are most extensive between Obigwe and Ogbidi (Picture 2), as well as near Obagi along the road to Ede. Here, the grasses grow to heights of 6ft to 10 ft with occasional stands of hydrophytes and tropophytes. This vegetation type may be referred to as an environmental pan climax especially as it is restricted to abandoned water courses silted up by alluvial and sandy deposits.

Drainage Characteristics and Flooding Problems

Picture 1 (Top) Swampforest in rainy season
(middle Ponds in dry season (Below) Bailing of pond

Areal Aspects of the Integrated Drainage Character

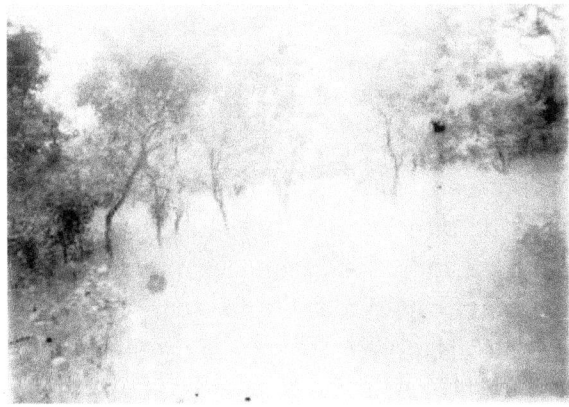

Picture 2 Grassland fresh water wetland along Obagi Ede road

The interaction of the vegetation and soil characteristics with the rainfall pattern creates an environment of drainage deficiency climaxed in soil saturation and flooding. The vegetal cover over the low relief of the area minimizes run off and evaporation. These promote stagnation of water on depressions. By promoting a thick soil cover, better soil texture and breaking up of the impact of rain drops on the land the vegetation contributes to the increase in the general infiltration rate of the soil. The water that has so infiltrated ultimately recharges the water table below. Consequently, this would tend to rise to very near the surface of the land in the well-drained sites. Otherwise, it generally occurs above the surface in the swampy and low lying localities making these areas remain under water throughout the rainy season (Picture 3).

Between June and October, (Figure 12) the long continued and steady rainfall (occasionally lasting for many days) results in so much saturation of the ground that through flow is deflected closer and closer to the surface. Consequently, the upper more permeable soil layers become filled up from their bases because through-flow is unable to carry away the water fast enough. In time, the soil becomes so saturated up to the surface that overland flow occurs. [35]

---

35 Chorley, (Ed.) op. cit., p. 219.

Drainage Characteristics and Flooding Problems

As a consequence of this phenomenon, the drainage channels which invariably are like closed basins ultimately overspill their banks and inundate or flood the over-flow plains and much of the adjoining lands. This is accentuated by the rise of water to and above the in-bank capacity of the scour routes and drainage channels following direct supply of water from precipitation falling into the depressions. Such inundation of over flow plains and abandoned drainage routes may last for some days or weeks before the water recedes. Figure 14 shows the profile of the topography of the Ubie-Oyiba area near Ahoada which illustrates the areas affected by rain flood.

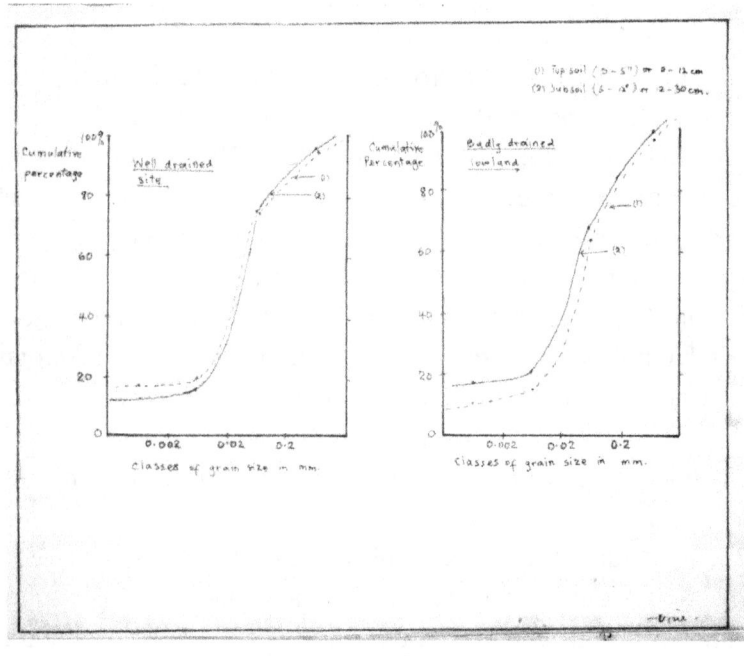

Figure 13 Soils samples Cummulative frequency curves in a location, Ede, in study area

Areal Aspects of the Integrated Drainage Character

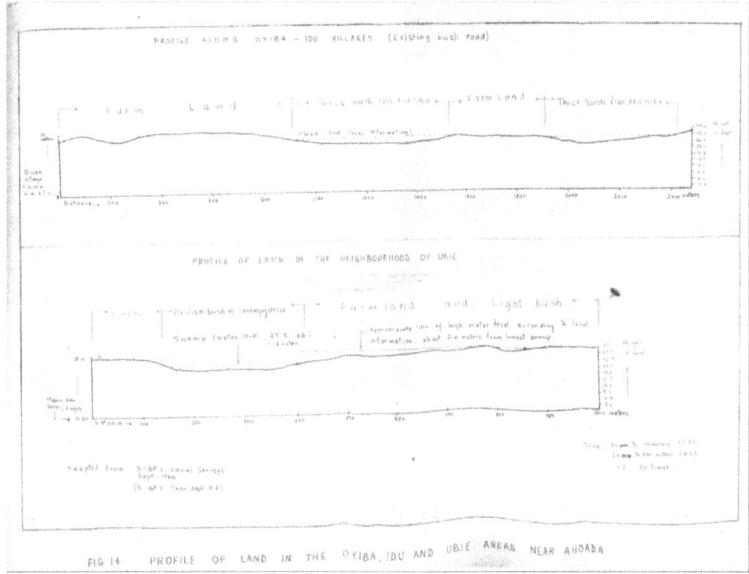

Figure 14 Land Profile in the Oyiba Idu and Ubie Areas

River Flood, ("miNi-iji, in the Ogba local parlance) or Flood due to the Rise of the Orashi Water Consequent upon the Niger Floods

Annually, towards the end of the rainy season but usually in October, River flood which depends principally on the character of the Niger River and Orashi River in terms of discharge, channel width, depth, and velocity relationships accentuates the problem of flooding and bad drainage in the Ahoada East and West/Ogba-Egbema Ndoni area. This is because the pattern and form of a river channel are determined and maintained by the dominant discharge of the channel, otherwise defined as elevation of water surface just below the stage at which the adjacent lands to the channel become inundated. For the Orashi and the Sombreiro, discharge is related to the b, f, and m exponents (Chapter 2 pp. 14-18). It was noticed that as discharge increases, width increases faster than depth in the Sombreiro; whereas in the Orashi, change in depth is slightly more than that of width. The m values of the two channels being 0.3 and 0.4 for the Sombreiro and Orashi, respectively indicate minimal velocity change when compared with the other exponents – b and f. Because of the small differences in the exponential value in relation to the constants as discussed previously (pp. 14-18) a high increase in discharge, for example discharge at high water stage, would so much alter the b, f, and m relationships with discharge that overbank flow would occur. This is because bank-full discharge is related to the high stage of the river – i.e., the stage at which the stream channels become filled and above which they overflow their banks.

As revealed in Figures 15 - 17 there is some degree of interrelationship among discharge hydrographs, water levels, and rainfall graphs for the Niger, Orashi and Sombreiro rivers at Onitsha, Okarki and Ahoada, respectively. The graphs of water levels (river stage) and rainfall show close relationships generally for the three rivers, but most significantly in the Sombreiro. This close relationship indicates that intense rainfall, which may induce a rise in river stage through direct precipitation falling on river channels, surface runoff and sub-surface inter-flow (moisture moving laterally through upper soil horizons), is a determinant of bankfull discharge and hence flooding in these rivers.

From the discharge graph for the Sombreiro at Ahoada (Fig. 15) one infers, therefore, that the Sombreiro draws its discharge mainly from local rains. This is because its discharge falls when that of the Niger is in full spate but increases and falls in consonance to the rainfall regimes of the area through which it traverses. But the width, depth, and discharge relationships for the Sombreiro ($w = 2.5 \, Q^{0.7}$; $d = 4 \, Q^{0.2}$) whose exponents are 0.7 and 0.2, respectively account for its development of a wide dish-shaped channel. When this is considered in relation to its small discharge values which range from 140 cusecs to 3,000 cusecs for a river mouth whose volume is $8,000 \times 10^6$ cu. ft[36] it becomes evident why there is hardly

---

36 Nedeco, Waters of the Niger Delta, p. 198,

such a significant rise in the volume of the river (its bankfull discharge) to cause flooding. Only the abandoned waterways adjacent to the channels are flooded by rain floods as earlier described. It is however, postulated that possibly during very high Niger discharge when the fresh water back-swamps get flooded through the Orashi flood (derived also from the Niger), some Niger water will find its way into the Sombreiro. This could result from direct flow through the interject creeks linking the Niger with the Orashi and the Orashi with the Sombreiro. Possibly, it could also be through sub-surface of the 8% yearly Niger discharge assumed to be temporarily stored in the Niger flood plain and swamps between Onitsha and Samabri.[37] Perhaps, this "connection" in drainage between the Sombreiro and the Niger brings about reminiscences of the earlier days when the Sombreiro was an active Niger branch.[38]

When the graphs for the Niger and the Orashi are compared (Figures 16 and 17) an intimate relationship between the discharge and stage in the River Niger is revealed. In the Orashi River as also in the Niger River, period of highest rainfall lags behind both highest water level (stage) and peak discharge periods. This is because flow of water contributing to the attainment of peak discharge in rivers is determined by the amount of water in storage. The water thus stored in the soil and bed rock over the basin as a result of infiltration and percolation from earlier rain will flow into the stream to re-charge the storm runoff in the river, usually after the period of maximum rain. It is only when the quantity of water drainage from the basin to the stream has reached a maximum that discharge reaches its peak.[39] This accounts for the occurrence of highest water level and peak discharge late in October when the rainfall is declining.

But comparing the graphs in Figures 16-18 with the diagram in Figure 19 indicating the "Orashi yearly discharge distribution as

---

37 Ibid., p. 104.
38 Ibid., p. 106.
39 Chorley, op. cit., p. 405.

percentage of Niger discharge up stream of Okarki" reveals that the Orashi discharges (bankfull discharges) are not mainly dependent on the rainfall of its drainage basin (Figure 11) as is the case with Sombreiro. Rather, it is dependent on the Niger discharges as the Okarki hydrograph (Figure 18) shows. This confirms the observation that during Niger discharges higher than 28ft between Onitsha and Aboh, part of the Niger water flows into the Orashi[40] to raise the latter's water level and hence dominant discharge.

Since the change in depth of the Orashi channel is just slightly more than the rate of change in width with increasing discharge (reference exponents a, c, k for the Orashi p. 15) the Orashi channel cannot deepen or widen fast enough to accommodate this increasing discharge. Moreover, an examination of the constants a, c, k for the Orashi already discussed (p. 15) reveals that the rate of width to discharge is the highest for all the parameter. Consequently, the Orashi is intrinsically characterized by lateral outflow within this section of its channel with the result that the adjacent lands become inundated during high water stages

The discharge relationship between the Niger and the Orashi river systems (Figure 18) validates the idea that the Orashi has a dual drainage character: being independent most of the year by drawing its discharge solely from its catchment area but discharges some of the Niger water during high Niger discharges. In this regard, the Orashi could be considered a part of the physical conduit system for the Niger. It should be remarked that the function of discharging some Niger water is affected primarily by adventive transfer through the now degraded Osemiri and Ndoni (Onita) creek and other bifurcating and obsolete channels of the flood plain. But the consequences of the dual drainage character of the Orashi in terms of its flooding potentiality are that if the river is completely cut off from the Niger, its discharges will fall significantly. Perhaps, bankfull discharge may be impossible to attain if it is robbed of

---

40   Nedeco, Waters of the Niger Delta, p. 104.

the 1.5%[41] (equivalent ⅓ of its discharge) Niger discharge that it receives yearly.

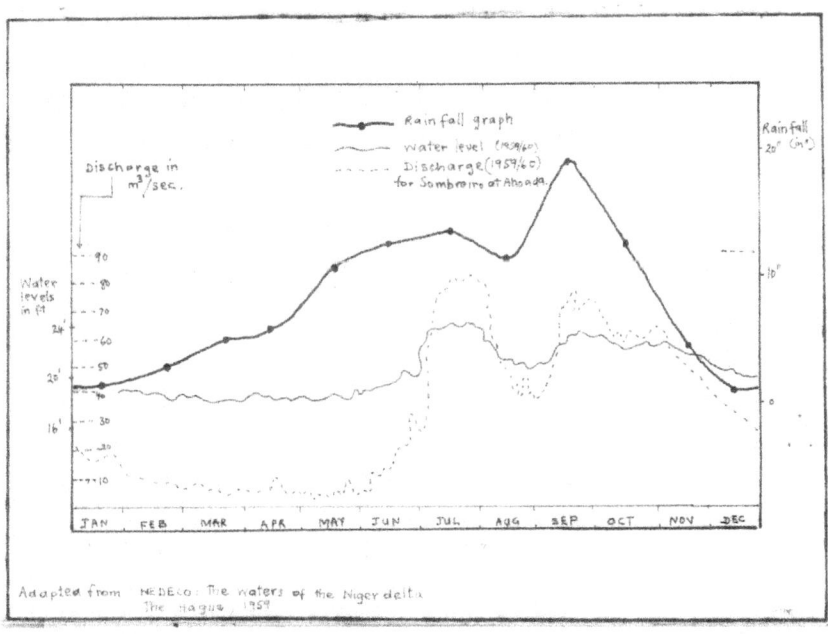

Figure 15 Somreiro Riverwater level and discharge and Rainfal for a year at Ahoada

---

41 Idem.

Drainage Characteristics and Flooding Problems

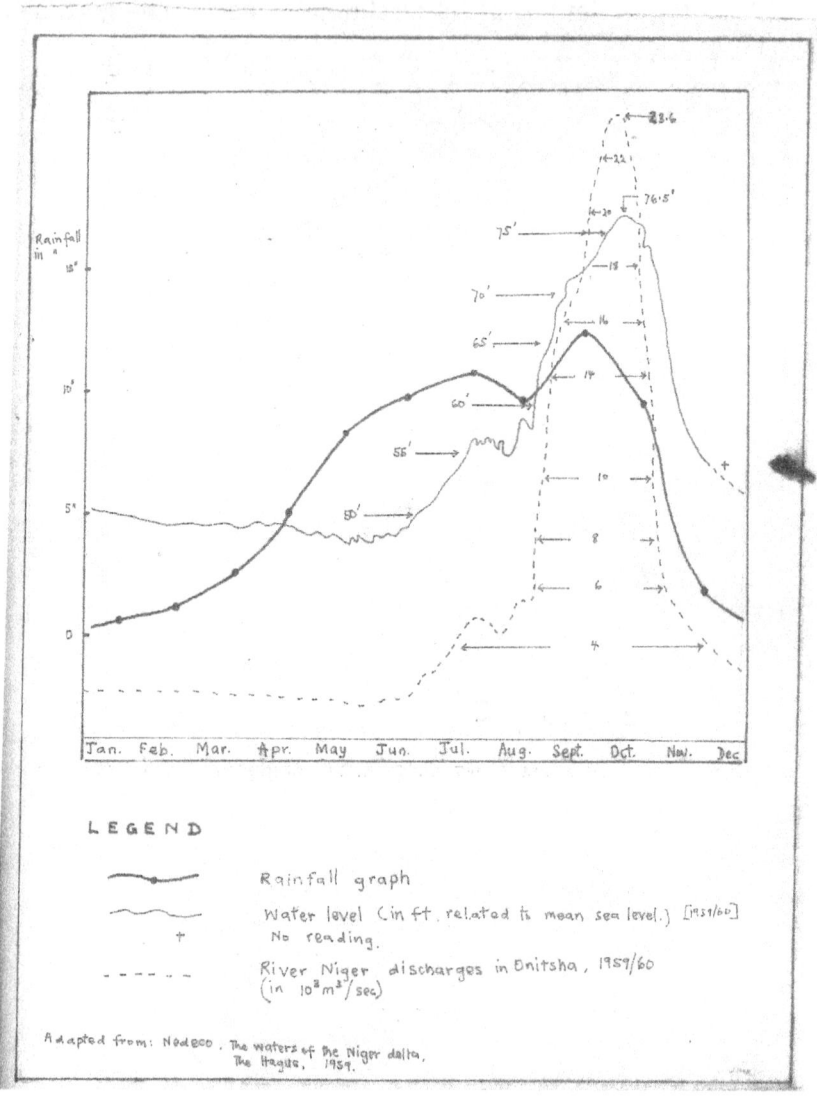

Figure 16 Niger River Water level and discharge and Rainfal for a year at Onitsha

Areal Aspects of the Integrated Drainage Character

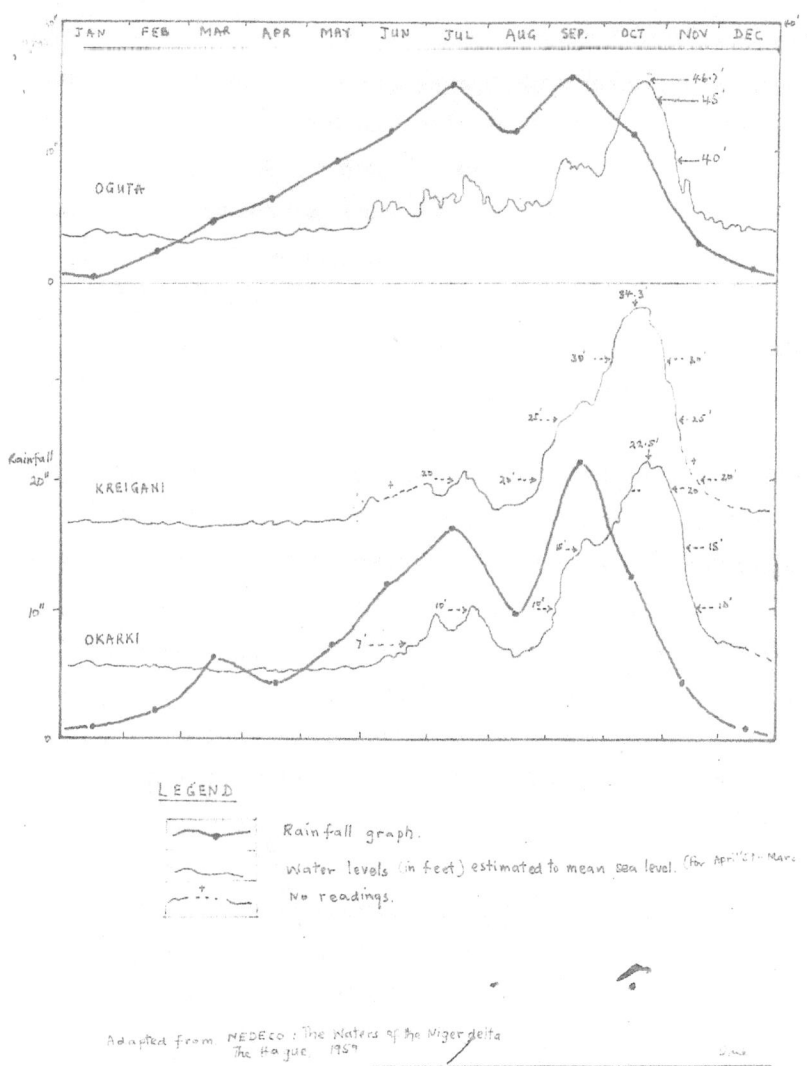

Figure 17 Orashi River water level at Kreigani and water level and rainfall for a year at Oguta and Okarki

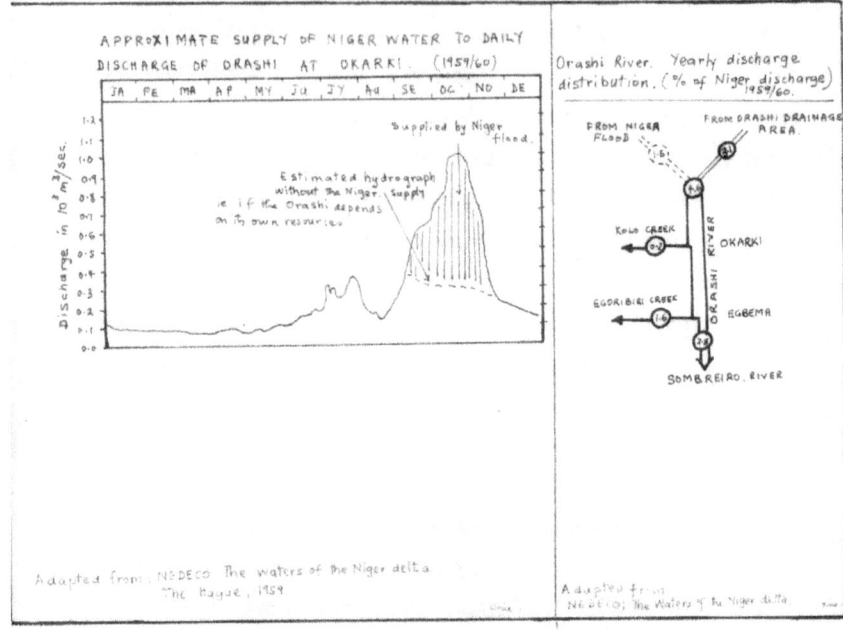

Figure 18 Supply of Niger water (discahrge) to Orashi River at Okarki
Figure 19 (chart on the right) Supply of water to Orashi and discharge distribution

Flood Probability and Frequency in the Orashi

The paucity of data on yearly flood discharges of the Orashi does not permit any statistical estimate of flood discharge, flood probability and frequency in the area of study. But data obtained from survey maps prepared by the Shell-BP Company and from random sources are tabulated below: (Table 5). Figures which show overbank flow of the Niger at Onitsha have been analyzed (see Appendix G) and used as basis for the graph appearing in Figure 20. The three year and eight year records for the Orashi and the Niger, respectively show that in each of the years on record the rivers attained a stage above their approximate water levels. This correlates with what is observed as a normal river where frequency of bank full flow has a recurrent interval averaging 1.5 years[42].

---

42   Leopold, et al., op. cit., p. 319.

Areal Aspects of the Integrated Drainage Character

TABLE 5: WATER LEVELS ON THE ORASHI AT KREIGANI AND OKARKU

Station	Average Water Level	Water Levels (ft)			Level of Flood Above Mean Water Level (ft)		
		1959	1960	1964	1959	1960	1964
Kreigani	28ft	34.3"	Figures unavailable	40	6.3ft	--	12 ft
Okarki	22ft	22.5	42	32	0.5ft	20"	10 ft

If the average water level on Orashi is taken to be approximately 28ft (representing bank-full stage) it means that the Orashi floods range between 6ft and 12 ft (for the years on record) above the mean water level. This corresponds with visual observations (e.g., water marks, water borne debris in the foliage of trees and reports from local people) which indicate 10 12 ft maximum flooding in the Orashi around Kreigani. Fig. 21 reveals the approximate flood level between Mbiama and Kreigani. It shows the general tendency for the floods to be highest towards the source of the river more than in the downstream section.

The difference is explainable by the increase in channel efficiency downstream where increasing discharge makes a corresponding increase in width (refer to the b, f, m values discussed, p. 16). Since the channel tends to widen, and there is also the entry of more distributaries, the flood is therefore, dissipated downstream.

Drainage Characteristics and Flooding Problems

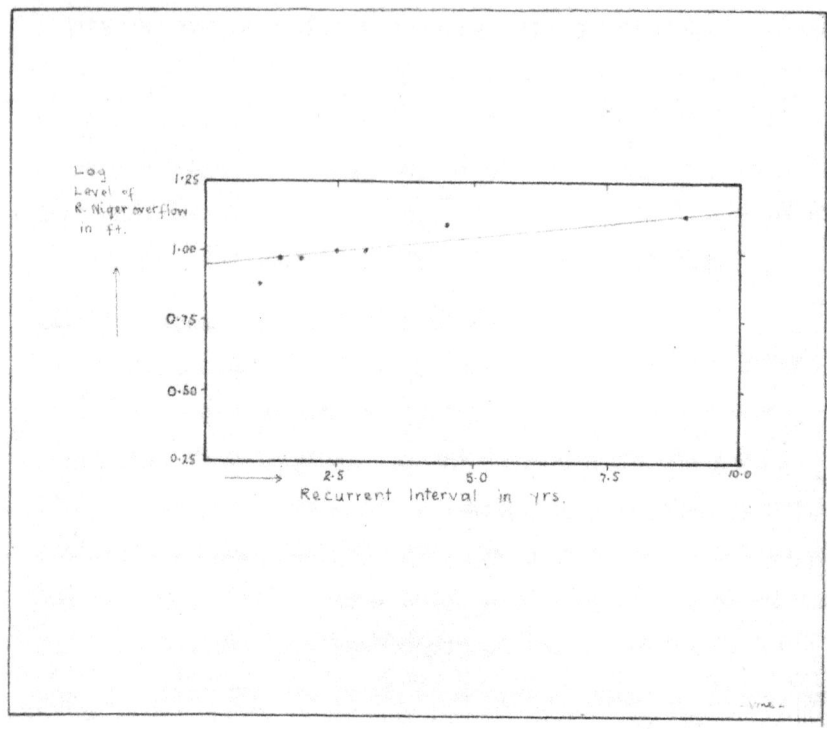

Figure 20 Frequency of River Niger Overflow at Onitsha

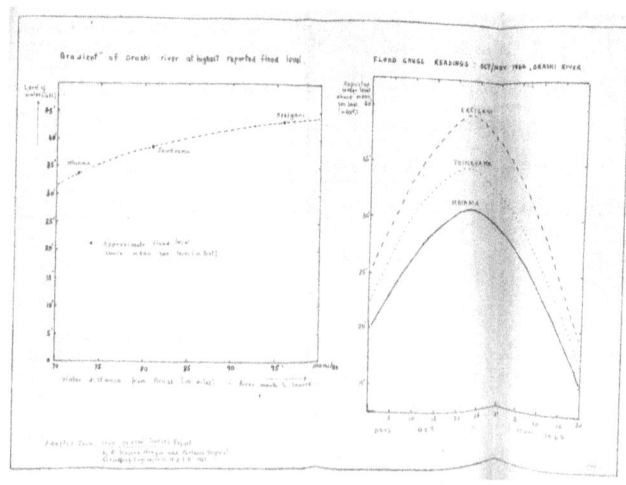

Figure 21 Orashi Flood levels at Mbiama, Joinkrama and Kreigani

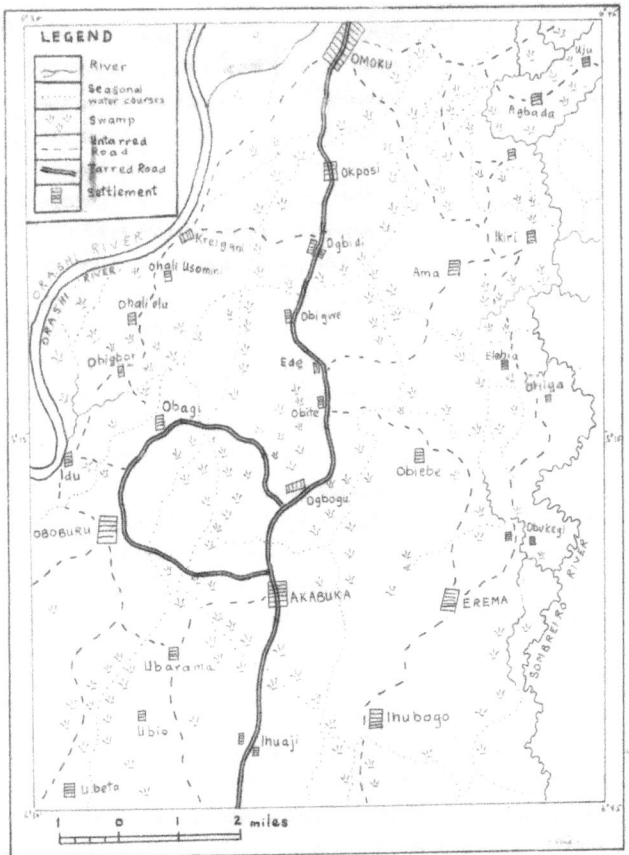

Figure 22 Settlements and some drainage features in central Orashi- Sombreiro Plain

The various parameters of the hydraulic geometry of the Orashi and the bio-climatic environmental factors of rainfall, soils, geology and vegetal cover are causally related factors responsible for the flooding phenomenon of the study area. The high yearly discharges of the Niger accentuate the Orashi discharge so much that the increase in channel size does not keep pace with the volume of incoming discharge. Consequently, there occurs overbank flow and flooding. While the mode of overbank flow may be infrequent, depending on the nature of rainfall and the Niger discharges, the interaction of the bioclimatic factors with the topographical disposition, lithological and structural characteristics of the environment influences the drainage by bringing about seasonal inundation of the land as rain floods. Although these physical factors discussed are general causal factors, flooding and bad drainage have been induced and facilitated by human interference on the landscape. So, it is the burden of the next chapter to investigate the contribution of human factors in the drainage problems of the communities of present Ahoada East and West and Ogba Ndoni Egbema Local Government Areas within the Orashi-Sombreiro plains.

4

Climate Change and Human Factors in the Drainage Characteristics of the Ahoada East and West and Ogba Ndoni Egbema Areas of the Orashi-Sombreiro Plains

As discussed in the previous chapter, the main factors which play a part in the drainage characteristics and especially flooding in the study area can be classified in two, namely land form and land use factors, and climatic factors. In this chapter, attention is focused on the contribution of climate change and human induced factors characterized by the pattern of land use and development in the drainage characteristics of the study area.

Climate Change Factors

For the past several decades, the world as a whole has been experiencing global climate change which manifests itself in extreme weather events including: extreme rainfall and flooding, higher temperature and violent storms. These extreme weather events are predicted to continue in intensity with major impacts on human activities such as agriculture and global food systems security, as well future population growth and people's quality of life

The intensity and frequency of the changing global climate and the impacts of this change are being observed more and more in Nigeria as in other parts of the world in recent decades. It has contributed to intensify the frequency of the flooding phenomenon observed in the Orashi-Sombreiro plains in recent decades. Within the past eight years (2012-2020), the study area has witnessed severe flooding events with disastrous effects on flora and fauna and on humans, more than in the past.

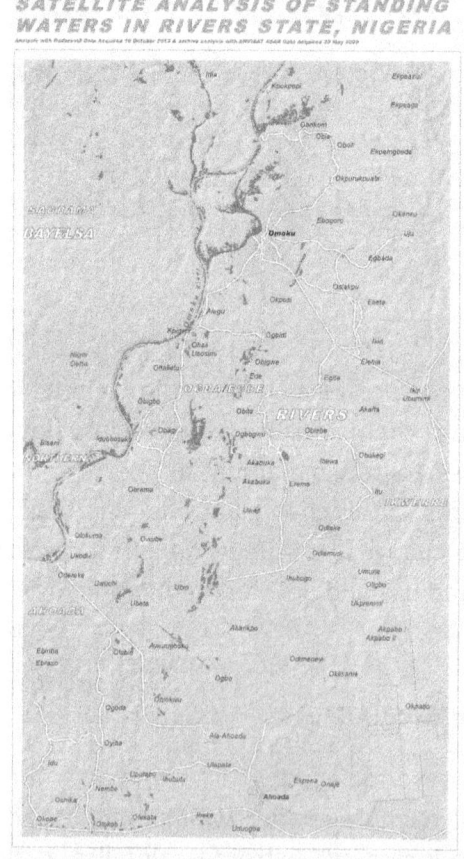

Figure 23 Flood Water in Ahoada East and West Communities during 2012 Flood

Human Factors

Apart from the physical climatic change factors, human activities have also contributed to the bad drainage and especially flooding phenomenon with very disastrous effects in the study area.

Although the study area is mostly rural but in recent decades, it has attracted a lot of people because of the economic potentials created by the oil industry. The increasing population of the area, which was estimated to be about 981,900 in 2016 with a population density of about 415 per sq. km., has put enormous pressures on land and water resources in order to produce sufficient food and meet other development needs of the region and the nation. The disturbance of the ecological systems especially the wetlands, due to oil production activities and the pressures on agricultural land use and the effects of local climate variability, especially in regards to rainfall regimes and intensity, are contributing to drainage and flooding problems in many communities. The problems are compounded because the area covered in this study has no integrated drainage system or land use plans to guide development activities. The available drainage network consists mainly of natural water courses and gutters constructed by people within their residences. Unlike in the past decades, land developers have tended to block natural water ways and build on wetlands without landscaping and creating channels for water to drain into the main natural reservoirs. In general, however, the main human factors causing flooding consist of:

- inadequate drainage network or systems in the communities and region in general
- blocking of existing natural and artificial drainage systems created by community efforts in the past
- filling up or construction of buildings or structures on natural water ways and fresh water routes without provisions made for rain water flow routes
- absence of well-connected and integrated gutters and canals system or other types of drainage channels from residential areas into ponds and reservoirs

Drainage Characteristics and Flooding Problems

It is a matter of common observation in the communities within the area of study, as in other parts of the world such as the Mississippi Basin (USA) and South-East Asia, that drainage handicaps (flooding and soil saturation) are frequently induced or accentuated by the action of human beings who unfortunately have to bear the burden of the problems.

The most obvious way in which man has contributed to the bad drainage character of the Ahoada/Ekpeye Ogba Egbema Ndoni area lies in the method of land use, Over this area, pits have been indiscriminately dug in both low lying area and easily drained sites to serve as fish ponds or borrow pits for earth for building of houses and infilling materials for lowlands and swamps across which route-ways are made. Rain water accumulating in these pits fills them up quite easily and very quickly especially in the swampy areas where the pits are shallow (4-5 ft deep). Consequently, the surrounding lands become inundated.

In the dry season, when most of this scour routes are dry, only the ponds contain water (Picture 1). But when they are bailed for fish (Pictures 1 and 2) debris and logs previously dumped into the ponds are thrown up. These obstruct the drainage channels leading water into the ponds. The practice of dumping new materials into the ponds contribute to the decrease of their water containing capacity. During the rainy season, the debris in these scour routes obstruct water-flow and so drainage in the area becomes impeded.

Excavations: The excavations made near built-up areas with the pile of earth obtained from the pits also act as barriers and therefore cause ponding after heavy rains. This is common in the sites of the borrow pits which supplied earth for raising the land level in the lowlands across which new roads have been made. The irregular surfaces thus created are favorable to inundation and deficient drainage. Some of these pits have been known to contain water all year round.

Construction Works: In many communities, buildings are constructed blocking natural water ways or wetlands, thus impeding flow of rain water. Also, construction trenches dispose of sewage from the Kreigani oil mill directly into the Orashi river which promotes the spread of flood into Kreigani town and the inland areas. The water pipe from the river to the oil mill also runs through this gutter which deepens towards the river. So, each time the Orashi river rises up to the level of this gutter, flood water moves through it onto the "backlands" of the river, inundating the oil mill, the market square and almost the entire village (Pictures 3 to 9). From here, the flood spreads to the numerous side creeks and swamps in the plain. Since the opening of the oil mill, this gutter has become the foremost distributor of the Orashi floods in the Kreigani oil mill area. This is because it is deeper than the other side valleys in the area. Once the Orashi river's overflow is added to the intense rainfall, communities along that axis, such as Al Igu, Ohali, Obigbor Obagi, and Oboburu tend to be flooded, too.

Picture 1 (Top) (a) and (b) Houses affected by Orashi Flood in Kreigani in the 1960s

(Bottom) Flooded section of Pioneer oil Mill at Kreigani showing level of flood, 1954

Picture 4 Flood level mark in Kreigani and the Pioneer Oil Mill Kreigani 1959

Drainage Characteristics and Flooding Problems

Pictures 2 - 4 Unpaved roads flooded after heavy rain

Picture 5 A perennial pond with water hyacinth in center of Oboburu town

Picture 6 Flood scenes in some communities of the study area

Drainage Characteristics and Flooding Problems

Flooded Communities in Ahoada West, Nigeria

Over 200,000 residents of Omoku live in fear over persistent flood

Flood: We're submerged, abandoned

Picture 7 Ahoada West Flooded communities

Bad drainage and Flooding Effects

Although climate change is seen as a contributor to the current intensification and frequency of flooding in the study area, flood damage of various types causes harmful effects on humans, their health and property as well as on public infrastructure, which may include roads, industrial and commercial businesses, and public services and facilities, as well as the economy. In essence, flooding can be seen to have both positive and negative effects for society. When flooding is managed properly, the excess water can be harnessed to support fishing and crop farming. In Nigeria or the study area in particular, the negative effects in the long run may impact farming activities and crop production which may lead eventually to food insecurity, if there are no immediate plans to counteract the factors associated with climate change and weather variability. Apart from agricultural production, climate change impacts are felt on health, biodiversity as well as social and economic conditions. It affects people and the environment in general, as in both developed and other developing countries, as Nigeria.

Bad drainage and flooding have created a number of problems for the people of the study region. First, and perhaps the most important in terms of damages, are those caused by flooding. These effects include:

Flooding of farms and homes;
Disruption of communities and residents' activities;
Displacement and dislocation of people (homes, villages, town health centers and schools);
Destruction of fauna and flora, especially special rare species, e.g. pythons, sea cows; and
Disruption of oil industry operations and other economic activities of affected communities.

Although there is paucity of data recorded yearly to show damages done by the yearly floods, interviews made during the study reveal that farms in low lying areas are usually devastated by flood each year

(Pictures 8 and 9). One could estimate that hundreds of tons of farm crops worth millions of naira (thousands of British pounds) are lost each year because of flooding in this area. In fact, a radio broadcast on the 1970 flood of the Orashi estimated the loss of property caused by the flood in the Engeni area of Ahoada at about £100,000 [43]

Besides damaging farms, the Orashi floods have been notorious in damaging buildings on levee tops. In Idu and Kreigani areas where the levees have been broken down because of human action, the Orashi floods have spread to destroy buildings.

This has necessitated rebuilding of houses almost every year. The constant rebuilding of houses invariably involves expenditure physically in terms of energy, and materially in terms of money, which within a time dimension could have been otherwise invested in other directions if the circumstances were to the contrary.

Damages done to roads by floods in the area have been equally widespread. During the rains, many road stretches are rendered useless e.g., Ebiriba-Oshobelle road, Ogbidi-Kreigani road and the Ede-Obagi road. The overall effect of the flooding of these land routes is that for about three months in the year, the roads will remain closed to almost all kinds of traffic. Under the circumstance, communications between villages become hindered and trade becomes stagnant. Furthermore, the cost of maintaining the roads soars remarkably. This is because the floods are usually turbulent and so wash off some of the bridges and planks placed on the marshy and muddy sections of the roads

A classic example of an economic concern affected by the yearly floods of the Ahoada area is the Pioneer Palm Oil Mill at Kreigani. Although flooding of roads hinders transportation of produce to and from the mill, the flooding of the entire mill is of far greater consequence because of the loss in revenue which it entails. Information obtained from the mill authorities about the 1954 flood, as well as reports from the local inhabitants of Kreigani reveal that the Orashi flood reaches a level of 12" to 18" in the mill. Besides the mill, the approach roads to the mill also get flooded to a level 2 1/2ft to 3ft. In consequence,

---

43 Ministry of Information Radio Announcements, Rivers State, 27th October, 1970.

farms, houses, and vegetable gardens are destroyed. The 1954 flood in particular resulted in folding up of the oil mill operations with the attendant laying off the mill personnel for about two months (September to October). Production in the mill then totaled about £1,000 per month with a net profit of £200.[44] It had a staff of 20.[45]

When the closure of the mill during the flood season is considered in terms of costs, a loss of about £400 net profit to the management becomes evident. Generally, the non-payment of salary to the workers and the slowed economic activity of the period resulted in a shortage of money for the rural population which depends on the oil mill for quick sales of their palm produce. Furthermore, the flooding of the mill brought with it, rusting of the mill machines some of which consequently required replacement or re-servicing .

Waterlogging and soil saturation which occur in restricted areas, such as on the fertile lands adjoining the swamps, are responsible for the limitation of cultivable areas and the limited period of cultivation. The swamps and network of drainage channels also limit settlement expansion. The sporadic occurrence of these swamps, over the plains, accounts for the nucleated small village settlements pattern of Ede and Ogbogu. Ogbogu at this moment cannot expand except lineally along the Obite and Akabuka roads, because of the network of swamps and drainage channels surrounding it.

The occurrence of unwanted pools of water such as the "Atata" and "Ugbanikiro" in Oboburu and Ogbogu, respectively (Picture 5) has baneful effects on the health of the people.

They provide free breeding grounds for mosquitoes and other human and animal pests. Secondly, the use of their water by the rural population for domestic purposes facilitates the spread of skin diseases and worm infestation. Since most of the rural inhabitants depended on wells, swamps, and creeks for their water supply until recent decades, the drainage character of their area helped to increase the impurity of these sources. The **wells** became very shallow in the rainy

---

44   Figures supplied by Mr. J. Izidor, the Kreigani Pioneer Palm Oil Mill Manager, 1952-1954
45   Idem.

season, while the swamps and creeks act as depositories of impurities washed from the built-up areas. This condition promotes the spread of water borne diseases.

THE IMPACTS OF FLOODING IN RIVERS STATE, NIGERIA
(October 2012)

Hundreds of deaths
Hundreds of thousands displaced,
Hunger,
Thousands of houses, cars, farms, and schools destroyed
Transportation Routes washed away
Fishing and farming businesses grounded to a halt

Some Examples of Flood Impacts in Study Area (2012) in the words of the victims:

2012 FLOOD IMPACTS IN STUDY AREA

(i) *Dr. Jacob Egba, who hails from Okogbue in Ahoada East Local Government Area of Rivers, is one of the victims of flood that recently ravaged the area. Egba and his three children now reside in a camp set up for displaced persons by the state government. As the floodwater is fast receding, Egba is not so much worried about life in the camp but he is visibly troubled about how he and his family would be able to cope with life after their time at the camp. Egba's house and farm have been destroyed by the flood and he has repeatedly expressed concern about his coming phase of life when relief materials from the government, philanthropists and donor agencies would cease.*

(ii) *Mrs. Roseline Ogwe from Obigwe, Ogba/Egbema Ndoni Local Government Area of Rivers, also lives at the camp set up for displaced persons. Ogwe, a widow with three children, eeked out a living from proceeds from her cassava and cocoyam farms, prior to the flooding. "I lived in an old house left for us by my father and that house has been washed away; the little money I realized from the sales of my cassava and cocoyam had been spent here in the camp because the aid is not regular," Ogwe says.*

(iii) *Another widow, Mrs. Ellami Philip from Abua town, said she managed to escape with her six-year-old grandson from being trapped by the floodwater through the help of some youths who ferried them across to safety. "I did not have any time to harvest the little crops I have in the farm; everything has been washed away. I have not gone back to see what has happened to my house, box of clothes and other belongings." "I am seeing an entirely different lifestyle here, but my worry is what happens after now. We will not stay in this camp forever; how do we start again?" she asked.*

Egba, Roseline and Philip are just some of the several flood victims who are already becoming apprehensive about their survival when they eventually leave the displaced persons' camps. Such fears have elicited the concern of some observers who believe that government should urgently devise strategies on how to help the flood victims to cope with life after their stay at the camps.

(iv) *Children at the make shift camps for displaced flood victims at the Ahoada East/West and Ogba/Egbema/Ndoni local government areas are now exposed to diarrhea and other diseases. Our correspondent who visited some camps set up by the Rivers State government reports that a good number of children are not in good health because of the hazardous environment and lack of medical facilities. Some mothers who spoke to our correspondent complained that most of the children have been affected by diarrhea and malaria illness has become a trend among the children. The women also complained that the Ede-Oha camp in Ahoada East received insufficient mosquito nets and little or no medical care because the medical officers assigned to the camp have left (as at the time of filing this report) because of lack of medical equipment like drugs and others.*

An officer of a civil society group, Earth Skin Conservation Foundation also at the Ede-Oha camp who confirmed the death of two persons at the camp as a result of heavy and constant stooling of one of the dead victims explained that the hazardous environment of the camp has resulted to some illness on the campers and diarrhea on the children. "[T]he environment here is not friendly to the campers. Even as we speak u can see mosquitoes perching n biting us, u can imagine the experience at night. one of the 2 deaths recorded so far in this camp is as a result of heavy and continuous stooling which led his death while the other died of trauma," he said.

The N.G.O. officer also complained about the lack of cooking utensils and spices which has resulted to the campers having late breakfast as late as one p.m.

Flood Ravaged RiverCommunities
By **Fyneface Aaron** -
August 14, 2017

Rain flood has wreaked havoc in Egi land and its neighbouring Ahoada communities, following five days of heavy down pour which led to destruction of household property, livestocks, economic crops and buildings. This has compounded the devastating effects of cultists' activities in the entire clans, as majority people who had left home on self-exile for safety of their lives would have more sad stories to tell. No zone is left out in the disaster, starting from Umu-Obor communities of Oboburu, Ohali-Elu, Obigbo, and Obagi to Akabuka, Ogbogu, Obite and Ede in Etiti-Egi. The story is not different from Erema, Itu, Ibewa Obukegi in Ahiawhor zone with the cry of agony also coming from Obiyebe, Egita, Akabta and Obiosimini in Uso-Ozimini zone. The Tide while going round the communities, witnessed serious cracks on the walls of mud houses, especially in Itu, Erema, Akabuka and Ogbogu, while those living in block houses were seen bailing water from their parlours and rooms. A community leader and National Auditor two of Apex Egi socio-cultural organization, the Egi People's Assembly (EPA), Chief Nathaniel Oriji, said the effect of the rain flood could be compared to that of natural flood of 2012. "All we laboured to plant in our farms have all been destroyed. Our farms are like river, in few days time the food crops would all wither leading to another hungry season, which we have just survived," and attributed the uncommon rain flood this year to blockage of natural water channels by oil companies."

The chairman of Itu-Ogba Community Development Committee, Comrade Benard Obo, attributed the flooding of his community to a road constructed by the Niger Delta Development Commission (NDDC), which has no single drainage. According to him, during the construction, the community had appealed to the contractors to include drainages, but they told us we should approach NDDC because that was not included in their job specifications, but all our letters to the commission fell on deaf ears. He, therefore, appealed to the new project director of NDDC to come to the aid of the community by constructing drains in the community's section of the road. The situation in Ahoada town, the headquarters of Ahoada East Local Government is as bad as that of Egi as most residents along Omoku road, Abuja Housing Estate, Odiemerenyi and Ekpena roads have all evacuated while at Ula-Ehuda, the torrent from the community through the market was on high frequency. Schools, hotels, and churches were also not spared as people are now appealing to NEMA and the state government to come to their rescue. However, an environmental officer, and a town planner attached to Ahoada East Local Government council who pleaded anonymity said; "we are not authorized to speak to the press but the cause of the flooding is part of the climate change, we have been preaching," the Town planner said "most developers ignored our advice against building along water routes, today they are reaping the reward of disobedience, especial those living in the Ahoada new sites."

Flood Affects 10,000 In Rivers LGA
By ANAYO ONUKWUGHA, PORT HARCOURT
September 26, 2018

No fewer than 10,000 people in Ogba/Egbema/Ndoni local government area of Rivers State have been affected by flood currently ravaging the communities following the overflowing of River Niger and its tributary, Orashi River. Chairman of Ogba/Egbema/Ndoni local government council, Hon. Ifeanyi Odili, who disclosed this yesterday while speaking to newsmen at Omoku, the council headquarters, said no fewer than 2,000 persons have been rendered homeless. Odili described the situation as devastating, saying there was need for agencies of the federal government to come to the aid of the local

> government council. He said: *"The situation is very devastating. If you go to Ndoni district, precisely Utu, Utuechi, Ogbogene, Asaaga, Agwe, Omuikwu and parts of Ndoni Town, Agbaja, Isukwa and so many communities, the flood has sacked all of them.* "The people in the community live in just very little piece of land. We are trying to see how we can relocate them to other places like Ndoni Town, though, part of Ndoni Town has been overtaken by flood. "Part of Omoku, Okwuzi and Agga communities have also been taken over by the flood." He said that makeshift homes have been created by the council to shelter some of the residents now displaced by the flood. Odili said: "We are calling of the state government and the federal government to come and assist our people because they have been displaced. "Majority of them are predominantly farmers. Right now, they don't have anything to eat. So, they need food, they need foams, good medication and good water to drink."
> LEADERSHIP Nigeria Newspapers. Contact: editor@leadership.ng

> Flooding of farms forces Ede, Obite, Ogbogu, Obagi and other ONELGA communities to embark upon preemptive harvesting of premature cassava and oter crops (Dukuma, A. "The People's Herald", October 13, 2020)

One can make an endless catalogue of the effects of the drainage character of the area on human life, but it is evident from what has been said that the bad drainage character of this area contributes to the psychological and physical "idleness" and economic loss of most of the people in the rural communities of the study area during the wet period of the year.

Although it is difficult at present to assess quantitatively, the amount of damage caused by bad drainage in the area, observable evidences show that the flood phenomenon of the area as well as other related drainage problems in their wide and deep ramifications dictate considerably the pattern of socio-economic activities of the people.

Picture 8 Some 2018 Flood Impacts in study area

Flood Affects in Some ALI Ogba Communities, October 2020

Picture 9 Some 2020 Flood Impacts in ONELGA Ahoada West

5

Suggested Measures for the Improvement of Drainage and Mitigating Flooding Impacts in the Study Area

The drainage character of the Ahoada East/West and Ogba, Ndoni and Egbema areas covered in this study as explained in the preceding sections, are influenced by a great complexity of interrelated climatic, geomorphic and human factors variedly linked with the physical environment. These make the improvement of the drainage an uphill task, not only in terms of financial costs and human resources but also in the complexity and durability of any measure that may be undertaken.

Flood Risks Management and Mitigation Measures

Many countries of the World, both the advanced and less advanced, have put in place measures to reduce the incidence of flooding and its effects in their jurisdictions. In many of these places, there are two basic approaches adopted towards minimizing flood hazards in the local community. The first focuses on reduction of the probability of flooding which consists of measures that reduce exposure and increase adaptation, coping and recovery capacity. This is achieved

Suggested Measures for the Improvement of Drainage and Mitigating Flooding Impacts

through the use of the tools of environmental management especially land use planning and zoning. The second approach consists of measures that focus on reduction of potential flood consequences or effects. This approach comprises measures that help to reduce flood hazards and increase capacity of river or stream channels through flood plains protection and construction of dykes or embankments, as well as channels or canals. These are achieved through small and large scale engineering works.

As had been mentioned, the most spectacular aspect of bad drainages characteristic of the study area is flooding. In order to minimize the extent of this phenomenon or prevent its devastating effects on land beyond the immediate surroundings of the Orashi, large reservoirs and levees could be constructed. The natural levees bordering the river channels should be raised at the sites where they have been severely dissected by years of human interference. Concrete embankments or flood walls could be built at such reaches like the Idu-Kreigani sector through which flood flows to inundate the swamps and scour routes in the central parts of the plain and from there into the communities. The channels of the side creeks which form distributaries of the Orashi could be deepened to give them a high in-bank capacity to accommodate the overspill of discharge of the Orashi. For the effective distribution of the Orashi flood water into the major reservoirs, some of the flow channels should be widened, deepened or re-aligned. This will reduce the chances and degree of flooding of the adjacent lands including the built up residential areas in the communities. Consideration should be given to possible canalization of the most affected built up parts of Egi communities, such as Obagi and Oboburu to mention a few. In order to carry out these activities, a comprehensive reconnaissance survey of the region and land surveying of the communities should be undertaken with the help of the government and oil companies.

The source of excess water in the soil and on depression in the Ahoada area is precipitation. This means that the remedy for the bad drainage character of this area must be through surface drainage- the mechanism for removal of excess rain in flat places where water moves slowly through the soil to provide for adequate drainage.[46] The surface drainage systems which could be applied in this area include the installation of enough field ditches to drain each individual pocket of stagnant or impounded water, the grading and smoothing of the land to provide a surface that will not interfere with distribution of surface water and the installation of field drains (gutters) to move excess rain into outlet ditches. Efforts should be made to elevate land routes at low lying parts of the land. This could be done by infilling of the depressions with earth as was done on the road stretch between Obigwe and Ogbidi. Here, embankments as high as 6-8 ft above the natural land-surface have been constructed. Concrete culverts have also been made across the road for easy passage of water from one side of the road to the other. Such measures could be applied on the Ngbede-Aga, Ubie-Oyiba road and Kreigani-Ogbidi road, where flooding of the roads is profound in the wet season.

Perhaps, one of the greatest drainage problems to combat apart from the periodic Orashi River flooding is that which is related to the numerous swamps or wetlands that occur in the area. The reclamation of these swamps may be done in a variety of ways. Firstly, the swamps could be cleared in the dry season to let sunlight reach the land surface and so ensure evaporation. This will facilitate the deepening of the ponds, construction of reservoirs and building up of the frontiers with earth materials. Secondly, the swamp basins could be developed into a network of lakes to serve for fishing and water-related sports to support tourism. Thirdly, new channels – canals – could be constructed to lead off water from the areas of bad drainage and stagnant pools into the reservoirs. Fourthly, attempts should be made to use the water which collects in this area for cultivation of crops on the frontiers during period of dryness.

Such crops like potatoes, groundnuts, vegetables, and rice could be cultivated almost at all seasons using water in the swamps through

---

46 K.V. Stewart, Jr. et al., *"Systems for Draining the Surface"*, Water, U.S. Year Book of Agriculture, p. 499.

a system of irrigation. Fifthly, a topographical survey of the region should be carried out to map out natural drainage lines. These drainage lines should subsequently be canalized in an integrated system to carry water to the main rivers especially the Orashi. The survey and engineering work entailed in this exercise will involve partnership with the Federal, State, and Local Governments, the Oil Companies and Niger Delta Development Commission (NDDC).

These measures obviously entail quite a lot of labor and capital, not to talk of time. But if they are pursued with determination, boldness, and imagination they will undoubtedly yield useful results in the reclamation of parts of the Ahoada area from the menace of bad drainage which characterizes it today. Moreover, they will create a favorable equilibrium of responses between "earth, water, and man". The government should be involved in the search for this equilibrium especially in measures for dealing with the frequent flood phenomenon and its impacts:

Many of the communities in the study area are rural and as a result have no land use plans or development control to guide the use of land. But they are characterized by intense exploitation of the natural resource base especially land. As a result, there exists:

- increased discharge of waste water that leads to road cutting and flooding;
- irregular and incoherent drainage pattern;
- water channeled from one plot flowing through other plots and causing damage and nuisance; and
- prohibitive cost of individual drainage control measures, and inadequate financial contribution to control measures by members of the community.

In the past, many of the towns and villages engaged in controlling street erosion and flooding by carrying out such activities as road filling, digging drainage for proper rain, waste water channelization, and construction of culverts over streams that cut across roads, through participation of community-based organizations such as the youth associations, age grades and help from the community at-large.

In order to address drainage and flooding problems in built up areas by local community efforts, consideration should be given to land use planning or landscape planning within the framework of physical planning of the communities. Landscape planning prescribes alternative spatial configurations of land uses, which are widely understood as a key factor in planning for sustainability. It deals with setting goals or courses of action for the beautification, functionality and ecological preservation of outdoor spaces – that is the built environment at-large. Also, it is an activity concerned with reconciling competing land uses while protecting natural processes and significant cultural and natural resources, reflective of human activity on land whose efficient arrangement and harmonious coordination are basic to physical planning with the objective of providing a pleasant, healthy physical environment for living, working, recreation and movement.

The Nigerian Urban and Regional Planning Law (1992) provides for preparation of town plans, rural area plans, local plans, subject plans and the control of development within their areas of jurisdiction, i.e., at the State and Local Government levels. Based on the above provisions, the local governments in the areas covered in the study should seek to develop a land use plan for the communities in partnership with the local communities and relevant State and Federal agencies and all the stake holders in the region. The development of such land use plans will serve as a basis for the communities to carry out their land development activities, control drainage/flooding and carryout landscaping activities in the communities.

The approach to adopt for effectively tackling the flood problem of the study area will require investment and taking actions where the greatest flood risks and benefits occur. In this regard, the central Orashi and Sombreiro plains should be considered a single flood area. This will require collaboration between various organizations, agencies, and communities and a focus on efforts to address flooding within the natural boundaries of the Orashi-Sombreio river systems rather than

Suggested Measures for the Improvement of Drainage and Mitigating Flooding Impacts

the political/administrative boundaries in the area. Also, structural and non-structural actions should be taken together to manage flood risks effectively aimed at achieving the following outcomes:

- Reduction in the number of people, houses, and properties at risk of flooding because of investment of public funds in actions that protect the areas at the greatest risks of flooding;
- Creation of rural and urban landscapes with space and structures to store and slow down the progress of floods;
- Integrated drainage systems that reduce flood risks and improve water environment;
- A well-informed public who understands flood risks and adopt actions to protect themselves and their properties and businesses from flood damage;
- Flood management actions that will harness excess water to support all year farming operations and other household activities in the area; and
- Flood management actions that will stand the test of time and be adaptable to future changes in the climate and the tempo of economic development in the communities.

Summary and Recommendations

Drainage characteristics in the Ahoada area are intrinsically related to the physical character of the Orashi-Sombreiro plains comprising essentially an intricate network of channels, numerous undrained depressions, and soils whose permeability varies with depth and a generally low relief. Within this environment, past and present fluvial processes of the Niger, Orashi and Sombreiro river systems, under conditions of high all year rainfall have operated and still operate in their interrelationships to create varying hydraulic processes. These carry with them widely ramifying drainage problems manifested by flooding and a confused and impeded drainage pattern, especially in the wet season. Moreover, the characteristic drainage of this area cannot be viewed as resulting from any one factor but from interplay of geomorphic, climatic, and human factors.

Although the drainage characteristics bring with them serious land use problems for the people, their influence on the pattern of socio-economic life in the area has been significant too. This is because it created an environment which the people have utilized for fishing, cultivation of raffia palms and for domestic water supply. Many of these environmental resource uses have been abandoned in recent times.

But the utility of the swamps, flooded lowlands and drainage channels cannot offset the negative effects which the overall bad drainage character of the area has on the people as manifested in soil saturation, flooding, marshy environment, and the development of many stagnant bodies of water. So, there is need for action to be taken to improve on the drainage by checking flooding or at least controlling the effects of its occurrence in the area. Because the swamps and the annually flooded lowlands constitute problems as well as pose a challenge to man's ingenuity in re-fashioning his landscape, efforts should be made to reclaim more of these for crop cultivation and settlement purposes.

Other actions that can be taken by the state and local governments should include provision of permanent shelters located at strategic centers similar to the internally displaced persons' center. Such centers can be easily activated during flood season to settle flood victims temporarily rather than leaving them on their own to deal with the emergency. Communities should also consider rain water harvesting to support crop cultivation during dry months of the year when water is scarce in some localities; which will allow for raising vegetable gardens and other crops during off-seasons to enhance food supply in the communities and region. Government should set up emergency funds to assist flood victims in rebuilding and catering for themselves during floods. It should also consider setting up flood alert and preparedness and monitoring systems for the region, so that people can be prepared before flood actually arrives to cause its disastrous effects in the area.

The local governments and the communities in the region should

Suggested Measures for the Improvement of Drainage and Mitigating Flooding Impacts

seek assistance of the Federal and State government, Niger Delta Development Commission (NDDC) and multi-national oil companies operating in the area to commission reconnaissance survey and land surveying of the region, as a whole and each of the communities for the purpose of constructing drainage channels (canals, culverts and bridges), as well as sumps in the appropriate areas or sites.

Indeed, it is by improved drainage and mitigating flooding impacts that the Ahoada and Ogba-Egbema and Ndoni areas will move towards socio-economic transformation of their landscape which the present century can offer to all who believe in the modern geographic concept that man can cooperate with nature to improve his environment.

Bibliography

1. **Allen, J.R.**, *"Quaternary of the Niger Delta and Adjacent Areas; Sedimentary Environment and Lithofacies,"* <u>Bulletin of the American Association of Petroleum Geologists</u>, May 1965.

2. **Chorley, R.J.** (Ed.), <u>Water, Earth and Man</u>: London, Methuen and Co., 1969.

3. **Hospers, J.,** *"Gravity Field and Structure of the Niger Delta, West Africa,"* <u>The Geological Association of American Bulletin</u>, Vol. 76 January – June, 1965.

4. **Information Unit**, Office of the Governor Rivers State *"The Geology of the Niger Delta and Surrounding Areas."* <u>The Oil Rich Rivers State</u> (N.d.)

5. **Leopold, L.B. et al.,** Fluvial Processes in Geomorphology, San Francisco, Wtt. Freeman and Co. 1964.

6. **NEDECO**, <u>River Studies: The Niger and Benue Report</u>, The Hague, 1959.

7. **NEDECO**, <u>River Studies: Water of the Niger Delta</u>, The Hague, 1959.

8. **Nwankwoala, H. O.** (2015) Hydrogeology and Groud water Resources of Nigeria, NY Science Journal, vol.1, issue 1 pp.89-100

9. **Prestrong, R.**, <u>Development of Drainage Patterns on Tidal Marshes</u>, Stanford University Publication, Geologic Sciences, X, 2, 1965.

10. **Short, K.C. & A.J. Stable**, *"Outline of the Geology of the Niger Delta,"* <u>Bulletin of the Association of American Petroleum Geologists</u>, L., 30 1966.

11. **Stewart, J.R., K.V. et al.**, *"System for Draining the Surface,"* <u>Water, U.S. Year Book of Agriculture</u>, U.S. Government Printing Office, Washington, 1965.

12. **Strahler, A.W.**, <u>Physical Geography</u> (2nd Ed.) New York, John Wiley and Sons, Inc., 1960.

13. **The Nigerian Urban and Regional Planning Law** (Decree 88, 1992)

Supplementary Sources

1. **Ayariga, David A**. *"Lan Use Planning as Tool for Flood Hazard Mitigation,"* (2014) B.Sc. Thesis (Kwame Nkrumah University of Science and Technology, Ghana

2. **Izidor, J,** A personal interview with Mr. Izidor, the Kreigani Palm Oil Mill Manager in 1952-1954 in the Office of the Production Manager, Rivers State Development Corporation in Port Harcourt, October 1970.

3. **Ministry of Information**, Rivers State, Radio Announcement, 27th October, 1970.

4. **SHELL - B.P., C.**, Topographical Dept. Locational Topographical Sheet, Ahoada North Location.

Appendix A: Fresh Water Discharge Measurements

1. ORASHI RIVER, BELOW OKARKI, GAUGE AT OKARKI

DATE	TOTAL DISCHARGE IN m³/sec.	v m/sec Av. VELOCITY	d WATER LEVEL IN Meters + m.s.l.	w CHANGE OF WIDTH WITH Q
4-1-59	110	0.38	1.7	145
8-10-59	900	0.82	6.6	170
2/23-11-59	120	0.42	2.4	170

2. SOMBREIRO RIVER. GAUGE AT AHOADA

3-7-59	68	.46	7.1	47
14-8-59	45	.46	6.5	30
8-10-59	40	.52	6.9	25
2-1-60	22	.4	5.9	23

Appendix B: Computed channel width, meander wave length and amplitude for the orashi and sombreiro

River	Estimated Channel Width (ft)	Meander Wave Length (ft)	Amplitude (ft)
ORASHI (a)			
(a) Ebocha	211.2	33792.2	4224
(b) Idu	502.4	22704	5491.2
(c) Mbiama	944.8	10982.4	5068.8
(d) Ukodu-Joinkrama	944.8	16896	7603.2
SOMBREIRO			
(a) At Ahoada	206	422.4	528
(b) Below Enema	100	1267.2	844.8
(c) Below Ahoada	180	844.8	1689.6

Appendix C: Stream Order and Stream Number in the Ahoada Area of the Orashi-Sombreiro Plains +

STREAM ORDER	STREAM NUMBER
1	30*
2	2
3	1

+ Computed from Drainage Map Fig 4.
*This number includes permanent and ephemeral streams/creeks.

Appendix D: Drainage Area and Cumulative channel length in the Orashi-Sombreiro Plains

DRAINAGE AREA (SQ. MILES)	CUMULATIVE CHANNEL LENGTH (MILES)
60	56
80	78
140	130
192	134
320	207
400	286
460	430

Drainage Characteristics and Flooding Problems

Appendix E: Climatic Data (Rainfall) for Ahoada

1. Mean monthly rainfall in inches for Ahoada based on records available and including 1957*

J	F	M	A	M	JU	JY	A	S	O	N	D	Y	# of Years
1.1	3.0	5.0	6.8	10.4	12.5	13.6	11.8	19.0	12.5	4.4	1.0	101.1	25

2. Number of Rainy Days Per Month for Ahoada

J	F	M	A	MY	JU	JY	A	S	O	N	D
2	2	5	5	10	15	20	20	25	20	5	2

3. Mean Daily Rainfall Intensity Per Rainy Day In The Month (in inches)

J	F	M	A	MY	JU	JY	A	S	O	N	D
0.55	1.5	1.0	1.56	1.08	0.83	0.68	0.58	0.76	0.63	0.88	0.3

4. Rainfall Variability for Ahoada for 20 Years 1941-1960

Year	1941	1942	1943	1944	1945	1946	1947	1948
Rainfall	86"	90"	79"	86"	100"	105"	82"	104"
Year	1949	1950	1951	1952	1953	1954	1955	1956
Rainfall	80"	82"	112"	105"	117"	92"	126"	103"
Year	1957	1958	1959	1960				
Rainfall	148"	99"	106"	114"				

Appendices

5. Climatic Data (Rainfall) for Oguta

# of Years	J	F	M	A	MY	JU	JY	A	S	O	N	D	YR.
7	0.3	2.3	4.8	6.9	9.9	11.2	15.2	11.0	15.5	11.1	2.9	1.0	92.3

6. Climate Data (Rainfall) for Okarki Area (Data – Based on Joinkrama Records)

# of Years	J	F	M	A	MY	JU	JY	A	S	O	N	D	YR
6	1.0	2.3	6.3	4.7	7.6	12.0	16.2	9.8	21.2	12.2	4.2	0.5	98.0

7. Climatic Data (Rainfall) for Onitsha

# of Years	J	F	M	A	MY	JU	JY	A	S	O	N	D	YR.
46	0.8	1.2	2.6	6.1	8.4	9.9	11.0	9.5	12.5	9.5	2.0	0.6	74.1

*From Nigerian Meteorological Service, Metrological Note No. 4, Mean Monthly Rainfall.

Appendix F: Granulometrical Composition of Soils**

1. Well Drained Site

    Depth of Soil Mechanical Analysis Percentage of Fine Earth

	Coars c. sand (0.2 – 2.00mm)	Fine Sand 0.02-0.2	Silt 0.002-0.02	Clay 0.002mm
Topsoil (0.5") Deep	20	63	2	13
Subsoil (5"-12")	20	56	2	18

2. Badly Drained Site

Topsoil (0.5")	33	49	4	11
Subsoil (5"-12")	30	47	4	17

**Adapted from Appendix A, samples taken around Provincial Agricultural Farm, Degema, and South of Ahoada in the Sombreiro-Warri Deltaic Plain. Anderson, B, Report on the Soils of the Niger Delta Special Area. Niger Delta Development Board, 1976, p. 37.

Appendix G: Annual Highest Water Level on the Niger at Onitsha*

Year	Highest River Stage or Water Level In Ft	Stage or Water Level Overflowing (i.e. Stage Above 28")
1950	35' 7"	7' 7"
1951	38' 9"	10' 9"
1952	35' 5"	9' 5"
1953	38' 0"	10'
1954	41' 8"	13' 8"
1955	41' 8"	13' 8"
1956	37' 10"	9' 10"
1957	41' 10"	13'

*Computed from Nedeco, River Studies, Niger & Benue Report, 1959, p. 361.

www.ingramcontent.com/pod-product-compliance
Lightning Source LLC
Chambersburg PA
CBHW070308230526
45470CB00002B/770